Geothermal Heat Pumps

Geothermal Heat Pumps

A Guide for Planning and Installing

Karl Ochsner

publishing for a sustainable future
London • Sterling, VA

First published by Earthscan in the UK and USA in 2008
Reprinted 2008

Copyright © Karl Ochsner, 2008

ISBN-13: 978-1-84407-406-8
Typeset by 4word Ltd, Bristol
Printed and bound in the UK by Cromwell Press, Trowbridge
Cover design by Yvonne Booth

For a full list of publications, please contact:

Earthscan
8–12 Camden High Street
London NW1 0JH, UK
Tel: +44 (0)20 7387 8558
Fax: +44 (0)20 7387 8998
Email: earthinfo@earthscan.co.uk
Web: **www.earthscan.co.uk**

22883 Quicksilver Drive, Sterling, VA 20166-2012, USA

Earthscan publishes in association with the International Institute for Environment
and Development

A catalogue record for this book is available from the British Library

The information and data contained in this book were produced to the best knowledge of
the author and were carefully reviewed by both the author and publisher. Nevertheless,
errors in the contents may exist. For this reason, data are given without the guarantee of
the author or publisher. The author and publisher take no responsibility for the existing
inaccuracies in content.

The German Library lists this publication in the German National Bibliography; detailed
bibliographical information is available on the Internet at http://dnb.ddb.de.

Library of Congress Cataloging-in-Publication Data
Ochsner, Karl.
    [Warmepumpen in der Heizungstechnik. English]
    Geothermal heat pumps: a guide for planning and installing / Karl Ochsner
        p. cm.
    ISBN-13: 978-184407-406-8 (pbk.)
    ISBN-10: 1-84407-406-4 (pbk.)
1. Ground source heat pump systems–Handbooks, manuals, etc. 2. Dwellings–Heating and
ventilation–Handbooks, manuals, etc. 3. Dwellings–Air conditioning–Handbooks, manuals,
etc.   I. Title.
    TH7417.5.O34 2007
    69–dc22

The paper used for this book is FSC-certified and totally chlorine-free. FSC (the Forest
Stewardship Council) is an international network promoting responsible management of
the world's forests.

Mixed Sources
Product group from well-managed
forests and other controlled sources
www.fsc.org Cert no. TT-COC-2082
© 1996 Forest Stewardship Council

FSC

Table of Contents

Preface

Since the release of the first edition in 2000, significant strides have been made regarding the acceptance of heat pump technology by operators and technicians, as well as by the public. Finally, the gap between national Kyoto Protocol goals and actual CO_2 emissions is being addressed with new policies. Heat pump technology, which has received too little attention until now, is now gaining some of the credit to which it is entitled. This is due to the fact that energy efficiency and environmental impact guidelines for heating technology are often best fulfilled through the use of heat pump technology.

This handbook is intended to act as a guideline for the planning and installation of heat pump heating systems. The contents are based on years of practical experience and results of ongoing development, as well as broad, international knowledge exchange.

Today, heat pump systems can provide an optimal indoor climate, optimal heating comfort, health and contentment in living and working areas. They do so more efficiently, more economically and with less environmental impact than most other heat sources (including some renewable sources). In addition, heat pumps can often offer the lowest heating, air conditioning and hot water preparation costs.

An honest and technologically up-to-date assessment of global emissions and economic factors favours heat pump technology. In the interest of energy security and operational reliability in the future, heat pumps deserve our attention today.

In this edition, the changes in energy prices and environmental requirements over the last 12 months are taken into account. The reader receives a newly revised handbook for a growing, innovative market, with new figures, graphics and tables.

Karl Ochsner, Dipl. Ing. ETH

Introduction

When Lord Kelvin described the theoretical basis for the heat pump in 1852, he could see no contemporary use for it as a heating device – but only for the provision of cooling in large colonial residences in Imperial India. He was prophetic in this respect, in that the use of heat pumps in buildings went on to be dominated by air conditioning and cooling. Millions of air conditioners, chillers and refrigerators (i.e. heat pumps) are manufactured and installed every year throughout the world.

The widespread use of heat pumps for heating buildings has taken longer to evolve. For many years, heat pumps for heating were seen as the domain of eccentric engineers who built and installed them in small numbers, primarily in Europe and the US. Over the last 50 years the technology has been slowly perfected, to the point where heat pumps for domestic heating are now mainstream technology in several European countries and in the US, and are growing in significant numbers elsewhere. It is difficult to find definitive numbers for every country, but at the five-yearly World Geothermal Congress in 2005 a worldwide review based on national submissions concluded that there are in excess of 1.3 million 12 thermal kilowatt (kWth) equivalent systems installed (Lund et al, 2005). These range in size from a few kilowatts (kW) to several megawatts (MW) of heating and/or cooling.

Despite his eminence in most things thermodynamic, what Lord Kelvin would probably not have foreseen was the role that heat pumps could play in reducing carbon emissions. In 1851 the world population was 1.1 billion. Today it has passed 6 billion, and is predicted to reach 9 billion by 2050. The use of fossil fuels to deliver heating in homes and offices is one of the largest sources of carbon dioxide (CO_2) emissions worldwide. With the release of the most recent report of the Intergovernmental Panel on Climate Change in February 2007 (IPCC, 2007), there is overwhelming evidence of the significance that manmade carbon emissions are having on global warming. Heat pumps are one of the few developed, reliable and widely available heating technologies that can deliver thermal comfort at either zero or greatly reduced carbon emissions. Most heat pumps are electrically driven, and as the generators reduce the amount of carbon dioxide emitted as a by-product of electricity production, so heat pumps will become even more carbon efficient with time – something that no fossil fuelled boilers can aspire to. Ultimately it is possible to envisage heat pumps being powered by zero carbon content electricity.

While it was not the original stimulation for the use of heat pumps, the reduction in carbon emissions arising from the heating of buildings is probably the most significant driver for their use in today's environment – certainly in central Europe. The International Energy Agency (IEA) Heat Pump Centre has recognized that heat pumps are one of the most significant single market available technologies that can offer large CO_2 reductions.

In addition, heat pumps offer other benefits: no on-site emissions, no on-site fuel storage, no flues or chimneys, a completely clean operation. Depending on local fuel tariffs, they can offer reduced heating costs, as well as deliver increased fuel price stability and fuel independence. With the arrival of Peak Oil, growing oil and gas demand, and the inevitable forthcoming rise in fossil fuel prices, the running cost advantages are likely to tip further in favour of heat pumps.

Because heat pumps derive the bulk of their delivered thermal energy from the environment (i.e. air, water or ground), this significant utilization of renewable energy is also capable of meeting the growing national and local targets for the increased adoption of renewable heat.

In the last half of the 20th century, heat pumps for heating, particularly in the domestic sector, evolved from a small, cottage/garage based operation populated by enthusiasts into a mainstream activity. Today, there are internationally known brand leaders capable of manufacturing tens of thousands of heating heat pumps every year – in addition to the many hundreds of thousands, if not millions, of heat pumps that are manufactured for cooling. The challenge now is to train enough knowledgeable sales staff, designers, installers and maintenance engineers to ensure that these heat pumps are correctly sold, specified, designed, installed and commissioned.

The challenge for anyone coming into this industry is that customers can rightly expect to be sold a fully functioning, modern, properly installed heat pump system. New players, whether individual installers or large heating contractors, are not going to have the luxury of learning and experimenting on the job over 10 or 20 years, at customers' expense. The experience of the established players, primarily in northern and central Europe and in the US must be adapted to local conditions quickly and appropriately so that high quality systems come on stream straight away. It is sobering to recognize that domestic heat pumps for heating, particularly ground source heat pumps, have had a history of poor quality selling, design, installation and equipment in several countries. So bad was this experience that the industry nearly collapsed before adequate training and standards were put in place to recover the situation. This must be avoided as the industry moves forwards in countries just embarking on the widespread adoption of heat pumps for heating.

One of the difficulties in acquiring the required knowledge is the lack of generic, English language reference material on heat pumps for the domestic sector. There is high quality material in the US primarily emanating from the International Ground Source Heat Pump Association (IGSHPA) – but much of this relates to water to air, and ground to air systems – which are very uncommon in the European domestic scene. High quality, non-English material exists in Europe, primarily in the form of manufacturers' technical guides – some of which have yet to be translated. Naturally they relate to the manufacturer's own product range.

In this book, a highly respected, long-term player in the European domestic heat pump sector has put together a wealth of information related to his experience over the last 30 years. While the author is himself a heat pump manufacturer, he strongly feels that generic information must be made widely available to all those wishing to get involved in this industry – in order to drive forward the number of quality installations. The first edition of this book, in German, was released in 2000. Since then it has been through four editions – and is here translated into English to meet the growth of heat pump installation, particularly outside of the well established German/Austrian/Swiss markets.

For those who already know about heat pumps, they can proceed directly to the main introduction. The focus of the book is on types of heat pumps and their selection and application. While aspects of ground design are referred to in relation to ground source heat pumps, this book is not intended as a detailed reference for would-be ground loop designers. The author pulls together his widespread, long-term experience in manufacturing and installing high quality water, ground and air source heat pumps. Combining this material with the relevant manufacturers' data sheets and technical guides should allow any would-be practitioner to embark on the design and installation of high quality heat pump systems that meet their customers' requirements and hopefully their expectations. Existing heat pump installers will undoubtedly find new and relevant information that will be of use to them as they expand their repertoire of installations.

For those who are new to heat pumps, particularly for heating in the domestic and small commercial sectors, a few words of explanation may be useful before delving further into the book.

Heat pumps do exactly what they say – they pump heat. Kelvin's great theoretical advance was that he overturned the notion that heat could only flow downhill, i.e. from hot to cold. The heat pump is able to collect low grade heat and deliver it at a higher temperature – albeit using some imported energy to do so. This is the principle of the refrigerator. The heating heat pumps discussed here collect low grade heat from the atmosphere (air), bodies of water (boreholes, lakes, rivers), or the ground. Using a refrigerant circuit, this heat is upgraded by an electrically driven compressor and can then be delivered at a useful temperature for heating. For cooling, the process is simply reversed: low temperature heat is collected from inside a building, upgraded and rejected to the atmosphere, water or ground. Using modern compressors and refrigerant cycles in well designed heat pump systems, it is possible to deliver heat at high energy efficiencies. A properly sized system using modern equipment can deliver between 2.5 and 4.5 units of heat (kWhth) for every unit of electricity consumed (kWhe). The ratio of these two numbers is commonly referred to as the Coefficient of Performance (COP). Readers should recognize that the quantity of electricity used to drive the heat pump is not insignificant – compressors are electrically demanding. However, depending on the fuel tariff for electricity vs fossil fuels, it is not uncommon to find that heat pumps can offer lower running costs than conventional fossil fuels or direct electric heating systems. Obviously it takes fuel to generate the electricity to drive the compressor. However, with modern power stations generating at ~35 per cent efficiency, and a heat pump with a COP of say 3.5, the heat pump will be 1.4 times more energy efficient than a gas fired boiler. With a modern combined cycle power station generating at say 45 per cent efficiency, and a heat pump in a new house with a COP of 4, the heat pump can be twice as energy efficient as the boiler. The cost savings will depend on the relative fuel tariffs. Depending on the fuel sources that are used to generate the electricity, the overall carbon savings can be very significant.

The reader may wonder why heat pumps are not the panacea for every building in the country. It is safe to say that any new building in the US or northern/central Europe that is built to the current building regulations is a suitable candidate for a heat pump. The total heating load and some, or all, of the hot water load can be met by today's heat pumps. There are some local limitations, for example in countries with single phase electrical supplies in the housing sector, that have upper limits on the size of compressor that can be used. The levels of insulation used in modern houses means that this is less of a restriction than it used to be. The main limitation on heat pumps arises in the older, existing housing market. This is because heat pumps currently deliver maximum temperatures in the range 55°C to 65°C. Traditional boiler systems can work at anywhere between 70°C and 90°C. In poorly insulated buildings, heat pumps may not be able to deliver year round comfort levels if such high temperatures are really required. In these situations, it is still possible to have hybrid – or bi-valent – heating systems where the heat pump works for as much of the heating season as possible, and the secondary source meets demands on the coldest days. This is not an unusual combination for air source heat pumps in mainland Europe. Improvements to the insulation, draft proofing and modification of the heat emitters to utilize the lower output temperatures of heat pumps is also an approach for older buildings.

In the context of this book, the types of heat pumps that are discussed derive their main heat input from the air, water or ground. While all three sources are used in Europe and the US, they tend to diverge in terms of the output side. The bulk of the domestic systems installed in North America deliver warm (or cool) air via ductwork into the building. In Europe, the most common form of heat distribution system in the domestic sector is water, either via wet radiator systems or wet underfloor (hydronic) heating. Many US units, particularly in the southern states, are reverse cycle units, capable of delivering warm or cool air. Thus in Europe the predominant heat pumps are air/ground/water-to-water heat pumps, whereas in the US and Canada they are more likely to be /air/ground/water-to-air.

For historical reasons the ground source heat pumps in Europe are often referred to as brine systems – because brine was originally used as the antifreeze in closed loop systems.

Heat pumps for heating and cooling buildings use the wider environment as their energy 'source' (or sink for cooling). Clearly the most widely available source is the air. The traditional difficulty for an air source heat pump is that on the days that the most heat is required, the outside air will be at its coldest – making it very difficult for the heat pump to achieve high efficiency, or possibly not being able to deliver enough output. With advances in compressor technology, heat exchanger design and control methods, this limitation is gradually being eroded and air source heat pumps for domestic heating are becoming more widely adopted. For well insulated buildings in moderate climates they can meet all heating needs: in more demanding situations they may have to work in conjunction with a secondary heat source.

Water and ground source heat pumps have a basic technical advantage over air source units in that water has a far higher heat carrying capacity than air, better heat transfer characteristics, and can be moved around very efficiently with small circulating pumps. Thus water source heat pumps have always been very efficient – it's just that most buildings have not had a suitable water source. Where there is a river, lake or borehole water, a water source heat pump is a very efficient means of delivering heating (and/or cooling). However, much of today's interest in heat pumps has been generated by the introduction of ground source (or 'geothermal') heat pumps. Technically these should be referred to as closed loop, ground source heat pumps. These use the ground around (or under) a building as their heat source or sink. By installing a suitably sized loop of pipework in the ground, water can be circulated to collect the renewable energy stored in the earth and deliver it to a water source heat pump. While very simple in concept, these closed loop systems are a little more complex in their design and operation, because the loop temperatures are engineered to change in order to induce the movement of heat by conduction through the ground. Considerably more care has to be taken in the selection, design and installation of closed loop ('geothermal' or 'brine') systems, compared to air and water source units, to ensure satisfactory, efficient operation. The designer needs to appreciate that closed loop ground source heat pumps are a coupled system comprising the building, the heat pump and the ground loops. All three need to be carefully matched to achieve a system that delivers the required amount of heat, efficiently, year-on-year for many years. While some general guidelines are provided in this book on ground loop sizing, it is recommended that designers of larger or more complex ground loop systems will need to garner further information on this specialist aspect. While trained heating engineers, plumbers or building contractors will be able to handle the installation of horizontal or trenched based ground loops, experienced drilling companies should be used to install borehole based systems.

While we have talked about the use of heat pumps for heating and cooling, it should be appreciated that heat pumps can also be used for generating domestic hot water – to varying temperatures. The simplest way of doing this may be to use an integrated stand alone air-to-water unit specifically designed for this purpose that comes fully integrated with a hot water tank. These are relatively new in the market place and have some way to go as a common method of generating hot water with heat pumps. The would-be practitioner should be warned that there are numerous suggestions for how a heating heat pump should be configured to generate hot water. Many thousands of systems are successfully delivering domestic hot water, but attention must be paid to the use of appropriately sized tanks, using significantly larger hot water coils or plate heat exchangers. Conventional hot water tanks used for boilers or solar panels will not perform satisfactorily on the lower temperature output from heat pumps.

There is a significant difference between the US and European approaches to hot water generation. Because of the dominance of water to air units in the US, it is common to use a specialized heat exchanger tapped into the refrigeration circuit – called a de-superheater. This can produce modest amounts of hot water while the heat pump is heating (or cooling) – but not at other times. In Europe, with air or water-to-water heat pumps, it is perfectly possible to switch the heat pump at any time to hot water generation. In particular, this allows the generation of hot water in the summer when the heat pump is not being used for heating. Installers must pay attention to local and national codes and practices related to domestic hot water systems, particularly in respect of pressurized systems.

As Lord Kelvin pointed out, heat pumps can be used for cooling – and in fact this application still dominates the world market in heat pumps. It is very easy for a heating heat pump to be fitted with a reversing valve to allow it to be used for what is termed 'active' (or refrigeration based) cooling. While there is a case to be made that ground source or water source heat pumps are one of the most efficient ways of providing cooling for buildings, there are reservations as to the widespread adoption of active cooling in countries where domestic cooling has not traditionally been required. It should be recognized that any form of active cooling can use significant quantities of electricity, and governments are concerned that unnecessary adoption of active cooling will wipe out hard won gains in energy conservation and carbon reduction. In climates that do not really require full-blown, active air conditioning, the use of passive cooling may be of benefit. This particularly applies to water and ground loop systems where simply circulating cool water from the source side of the heat pump through underfloor loops or fan coils can deliver a few degrees of cooling. This is referred to as passive cooling and is provided as an option with many European heat pumps. This allows the heat pumps to be optimized for the heating cycle, with no engineering compromise being required to deliver reverse cycle cooling.

With these basics in mind, the reader should now be in a position to appreciate the wealth of information contained in the body of this book. With the experience that the author brings, coupled to manufacturers' data sheets and technical guides, it should be possible to embark on the design and installation of successful domestic heat pump systems.

Robin H. Curtis
April 2007

References

IPCC (Intergovernmental Panel on Climate Change) (2007) 'Climate change 2007: The physical science basis – summary for policymakers', IPCC Secretariat, Geneva,

Lund J. W., Freeston D. H., Boyd T. L. (2005) 'World-wide direct uses of geothermal energy 2005', Proceedings of the World Geothermal Congress, Antalya, Turkey, April.

Thompson, W. (Lord Kelvin) (1985) 'On the economy of the heating or cooling of buildings by means of currents of air', Proceedings of the Royal Philosophical Society (Glasgow), vol. 3, pp. 269–272.

List of Acronyms and Abbreviations

ASSA	Austrian Solar and Space Agency
AWP	AWP Wärmepumpen GmbH (heat pump producer)
BW	Brine/Water
COP	Coefficient of Performance
EER	Energy Efficiency Ratio
EU	European Union
GMDW	Golf Midi/Maxi ground/direct expansion (type code)
GMSW	Golf Midi/Maxi brine/water (type code)
GMWW	Golf Midi/Maxi water/water (type code)
GWP	Global Warming Potential
HSPF	Heating Seasonal Performance Factor
IPCC	Intergovernmental Panel on Climate Change
ODP	Ozone Depletion Potential
RECS	Renewable Energy Certificate System
SEER	Seasonal Energy Efficiency Ratios
UN	United Nations
WW	Water/water

1

Reasons to Use a Heat Pump

1.1 Environmental benefits

1.1.1 Our environment is in danger

Over millions of years, the ancient forests and plants of our planet have produced the *oxygen* that we breathe today. The decaying plants and forests were swallowed up by the young Earth, and transformed over long periods of time into coal, oil and natural gas.

These are the fossil fuels that people burn today. During the burning process, the oxygen store is consumed and *carbon dioxide* is produced again. Another phenomenon is the enormous increase of methane gas in the atmosphere. There are several causes for this: the excessive production and use of natural gas world-wide

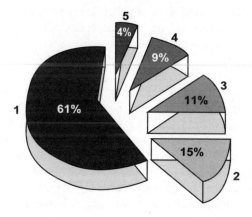

1 Carbon dioxide
2 Methane
3 Chlorofluorocarbons
4 Ground level ozone and upper atmospheric water vapour
5 Nitrous oxide

Figure 1.1 Emissions that contribute to the greenhouse effect and climate change
Source: Professor D. Schönwiese, Institut für Geophysik, Universität Frankfurt/M

has resulted in an increasing concentration of *methane* too. These gases intensify the natural *greenhouse effect* and present a *climate threat* (see Figure 1.1). In addition, the environmental pollution and damage caused by leaks in oil and gas pipelines and by tanker spills must be considered.

According to United Nations (UN) studies, the expected consequence is an average temperature increase of 1.5 to 6°C in the next century. This is predicted to give rise to dramatic climate changes: increasingly frequent storms, hail and heavy precipitation, as well as droughts and a rising sea level (see IPCC – Intergovernmental Panel on Climate Change – report released in February 2007).

The environmental pollution and damage caused by leaks in oil and gas pipelines and by tanker spills must also be considered.

1.1.2 Tracking down the culprit

Heating with fossil fuels is achieved mainly by burning oil and natural gas. During the chemical process of combustion, considerable amounts of sulphur dioxide, nitrogen oxides, soot and other pollutants are emitted, causing acid rain, damaging forests and endangering our *health*.

All types of combustion, including that with natural gas and biofuels, produces carbon dioxide (CO_2). This intensifies the greenhouse effect and leads to climate change. Heating an average family home using oil produces some 6000 kilogrammes (kg) CO_2 each year, and using natural gas some 4000kg CO_2. *Domestic heating* accounts for as much as 40 per cent of our CO_2 emissions in central Europe.

For this reason, new building codes are designed to limit primary energy consumption.

Heating with wood logs, wood chips or pellets is arguably CO_2-neutral provided that sustainable forestry is adopted, but still results in the emission of the pollution described above, as well as microparticles.

1.1.3 Heat pumps offer emission-free operation on-site

Heat pumps deliver heat without producing any on-site soot or other toxic exhaust. Depending on the heat source, heat pumps produce pollutant-free heating energy using solar energy, environmental energy, geothermal energy or waste heat.

So on the one hand you can store your white laundry in the heating room, creating another usable room, and on the other hand there are no pollutants which could harm your garden or backyard. Even your neighbours will be thankful for your commitment to national and global environmental protection (see Figure 1.2).

The leading heat pump manufacturers use only chlorine-free refrigerants with zero ozone depletion potential.

Depending on the mix of generating capacity on the grid, heat pumps generally offer excellent overall CO_2 emissions, i.e. as per the next figure with 50 per cent hydropower. It is also worth pointing out that even with 100 per cent modern fossil power generation there is an overall reduction in CO_2 emissions compared to individual condensing boilers, thanks to the efficiency of today's power plants.

Figure 1.3 shows how much on-site energy in kilowatt-hours per square meter per year must be expended. This energy is the quantity the user requires to heat the

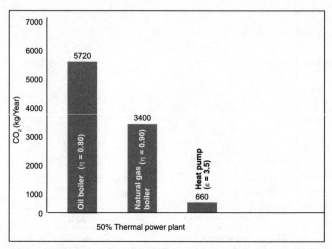

Figure 1.2 Emission comparison: typical single family house with 8.8kW heating demand
Note: Electricity: 50 per cent emission free (i.e. hydropower); 50 per cent thermal power plant.
Source: Institut für Wärmetechnik TU Graz, Energiebericht der österr. Bundesregierung 1990, actualised 12/1998.

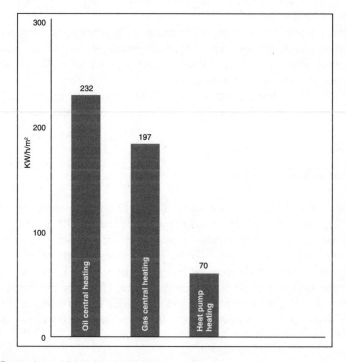

Figure 1.3 Energy demand for various heating systems
Source: GEMIS-VDEW

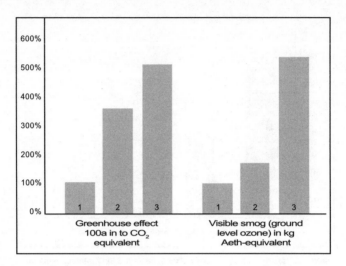

Figure 1.4 Comparison of environmental impact of heat pumps with oil or gas combustion
Note: Comparison of the impact of various environmental influences. 1. Heat pump (CH): Swiss electricity;
2. Gas: low-NOX, condensing boilers; 3. Oil: low-NO_x.
Source: AWP Zürich

required area (i.e. how many kilowatt-hours of electricity, litres of heating oil or cubic metres of natural gas).

One kilowatt-hour of electricity is equivalent to approximately 0.1 litres of oil. It is clear that heating with heat pumps requires significantly less energy than heating with gas or oil. The reason is that the heat pump draws up to 75 per cent of the required energy from its surroundings.

In Figure 1.4 the environmental impact of a heat pump is compared with that of gas and oil boilers. The CO_2 emissions from the (caloric) production of electricity are taken into account.

1.2 Operating costs

Depending on the efficiency of the heat pump, up to three-quarters of the required heating energy is drawn from the environment (without cost) when heating with a heat pump system. This environmental energy comes from the sun or in the case of geothermal systems with vertical loops from the ground. Heat pumps make use of the free renewable energy stored in air, water or the earth. With the help of a heat exchanger, the heat pump boosts the energy extracted from the environment to the temperature level required for heating. In Figure 1.5, a typical cost comparison (for Germany) is shown. With a heat pump, you can utilize solar energy economically year-round.

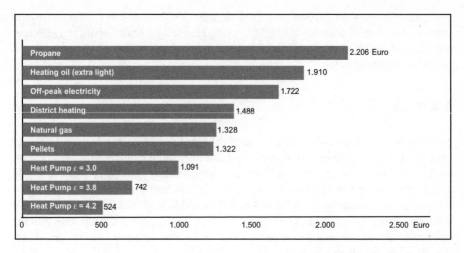

	Euro
Propane	2.206 Euro
Heating oil (extra light)	1.910
Off-peak electricity	1.722
District heating	1.488
Natural gas	1.328
Pellets	1.322
Heat Pump ε = 3.0	1.091
Heat Pump ε = 3.8	742
Heat Pump ε = 4.2	524

Figure 1.5 Operating costs per year, residential house 175m², heating capacity 9 kW, example Energie AG, 09/2006

1.3 Independence

The ambient energy sources of heat pumps – solar energy stored in the air, water and ground – all exist crisis-free, right outside your own door. Considering the forthcoming impacts of limited oil supply, and the rapidly increasing demand for oil

Oil Reservers 2005

Coverage in years

	Billion tons	Coverage
161.6 Bn. tons to world		41 Y.
36.0 Bn. tons to Saudi Arabia		68 Y.
18.0 Bn. tons to Iran		90 Y.
15.6 Bn. tons to Iraq		175 Y.
13.8 Bn. tons Kuwait		106 Y.
13.3 Bn. tons to Arab Emirates		102 Y.
10.8 Bn. tons to Venezuela		70 Y.
10.1 Bn. tons to Russia		22 Y.
5.3 Bn. tons to Lybia		66 Y.
4.9 Bn. tons to Nigeria		39 Y.
4.0 Bn. tons to USA		13 Y.
3.1 Bn. tons to China		17 Y.
2.2 Bn. tons to Mexico		12 Y.
1.2 Bn. tons to Norway		9 Y.

Billion tons

Figure 1.6 Oil reserves in billion tons – 2005
Source: Bundesanstalt für Geowissenschaften und Rohstoffe

and gas, it is easy to recognize that dangerous dependence on volatile foreign energy sources is an additional risk factor (see Figure 1.6). Because of their efficient use of local, ambient energy resources, heat pumps help to reduce dependence on imported fuel supplies.

1.4 Comfort

Heating with heat pumps offers the highest possible living comfort and ease of operation. The heat distribution systems commonly implemented with heat pumps, such as low temperature radiant floor and wall heating, guarantee a comfortable and healthy living climate. Low temperature radiant heat also minimizes overheating and excessive air and dust turbulence. Reversible heat pumps can also cool on demand during the summer. Heat pump heating systems generally operate quietly, automatically and are maintenance free. Fuel deliveries, disposal of ashes and chimney cleaning are all eliminated.

1.5 Security for the future

Heat pumps represent the *most modern* heating technology available. Today, heat pumps are no longer replacing just wood and coal heating, but also coke and central oil heating, and with increasing frequency even natural gas heating systems.

Additionally, there is the question of whether or not we will still be able to afford our heating systems in 20 years. The selection of a heating system should be a decision for decades. Future-oriented consumers will undoubtedly arrive at one conclusion: the heat pump. Even today, correctly implemented heat pumps can be the heating system with the *lowest operation costs*. With each increase in fossil fuel prices, the cost of heating with heat pumps becomes even more competitive when compared to oil, gas, or pellets. The savings will increase because with heat pumps, three-quarters of the energy remains free, even if electricity costs increase.

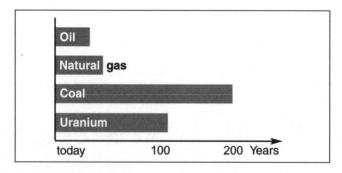

Figure 1.7 Supply of primary energy sources
Source: Ochsner

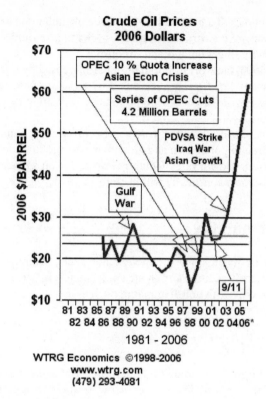

Figure 1.8 Development of oil price
Source: WTRG Economics

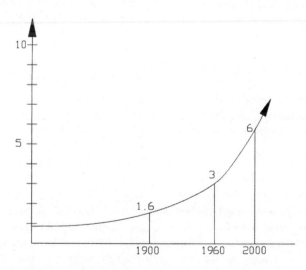

Figure 1.9 Growth of global population (in billions)
Source: Ochsner based on UN-Population Division 97

The energy resources of a heat pump are practically unlimited as far as quantity, availability and time are concerned. Figure 1.7 shows that resources such as oil and gas will become increasingly limited and expensive.

Since the year 1900, the world's population has increased by a factor of 3.5. The global energy demand, however, has increased by a factor of 10. Today, 6.2 billion people inhabit the Earth – by 2050 the population will grow to 9.1 billion (UN-Population Division, Average Assumptions).

In developing countries, the energy demand is growing at higher than proportional rates compared to population and gross domestic product (domestic, industry and trade). At current growth rates, China's demand for oil will exceed that of the US (currently the largest consumer) within 10 to 15 years.

1.6 Non-flammability

Heat pumps heat by means of a thermodynamic cycle, without combustion and flames. This significantly reduces any chance of a dangerous accident. Additionally, most units operate only with non-flammable refrigerants.

1.7 Responsibility for the future

We do not want to upset the ecological equilibrium which belongs to our children. Today, we carry the responsibility for tomorrow. In the future, oil and gas will be urgently needed as raw materials for applications in which they cannot be replaced. They are too precious to waste on domestic heating.

In addition, natural gas and oil imports burden the balance of payments of our national budget, while environmental energy represents a domestic asset.

1.8 Ideal for low energy houses

Heat pumps are the ideal heating system for low energy houses in which only a minimal heating capacity is required. Conventional heating systems are generally not available or technically or economically feasible for such low heating capacities.

1.9 Retrofit

Increasing energy costs, regulations or simply the breakdown of an existing boiler make a new investment necessary. Retrofitting a heat pump becomes more feasible as new technologies permit flow temperatures of 65°C. Furthermore, using air as a source is becoming more and more attractive as split units in particular offer a high coefficient of performance (COP). Air is available anywhere as a heat source.

1.10 Multiple functions

Heat pumps can also be used to cool in special configurations. With the proven increase in summer temperatures, cooling will soon be necessary in more and more geographical areas.

Heat pumps used for domestic hot water heating can also be used to *ventilate* or *cool* and *dehumidify* at the same time – without additional investment or operation costs.

Heat pumps used for controlled dwelling ventilation can provide additional heating or cooling using exhaust heat.

1.11 Public promotion

The installation of a heat pump is promoted by several authorities as an ecologically and economically important technology. This support varies from subsidies through the government, community and utilities to tax credits and reduced interest rates for credit financing.

1.12 Energy politics/laws

Heat pumps can be used in the most important end energy segments (see Figures 1.10 and 1.11), substituting for fossil fuels and reducing emissions.

Heat pumps are one of the few technologies that can lead to a significant reduction in CO_2 emissions and thereby help to reach Kyoto Protocol goals. For example, the German Energy Conservation Act, European Union (EU) Buildings Directive or the Australian Renewable Energy Certificate System (RECS) promote heat pumps through the placement of limits on primary energy consumption.

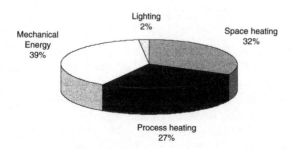

Figure 1.10 Space heating represents a significant portion of national energy consumption
Source: VEDW

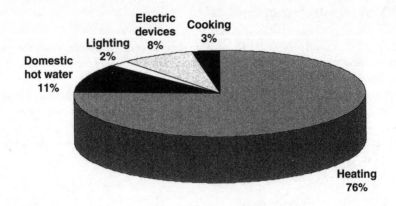

Figure 1.11 Heating and domestic hot water preparation are by far the largest energy consumers in private households
Source: VEDW

2

Theory of the Heat Pump

Principle of Thermodynamic Heating

The heat pump transforms thermal energy at a low temperature into thermal energy at a higher temperature which is suitable for heating purposes. This occurs in a closed-cycle process in which the working fluid is constantly undergoing a change of state (evaporation, compression, condensation and expansion).

The heat pump draws stored solar energy from its surroundings – the air, water, or the ground – and transfers this energy, plus the electrical energy used to operate the cycle, in the form of heat, into a heating or water heating circulation loop.

The Heat Pump Cycle

2.1 The principle

As shown in Figure 2.1 and Figure 2.2, the heat pump draws about three-quarters of the required heating energy from the environment.

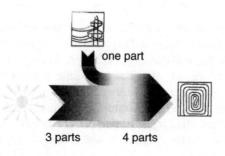

one part

3 parts   4 parts

Figure 2.1 Energy Flow Diagram
Source: Building Advice Guide

Environmental heat $^3/_4$ + Purchased energy, electrical $^1/_4$ = Usable heating energy $^4/_4$

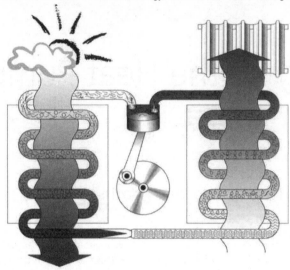

Figure 2.2 Principle of Heat Pump Operation
Source: BWP

2.2 The refrigeration cycle

Example of a heat pump refrigeration cycle with pressure and temperature values for refrigerant R 134a, 1.4 kW$_{th}$, A7/W50 (see Figure 2.3):

A7/W50 = input air at 7°C, output water at 50°C

2.3 Coefficient of performance

$$\varepsilon = COP = \frac{\text{Delivered heat energy (kW)}}{\text{electrical input to compressor (kW)}} = \frac{\text{Environmental Energy + el.i.to.c.}}{\text{el.i.to.c.}}$$

The coefficient of performance COP (ε) indicates the amount of delivered heat in relation to the drive power required. Therefore, a coefficient of performance of four means that the usable thermal output is quadruple the required electrical input. The coefficient of performance is an instantaneous value. It depends on the design of the heat pump and the operating characteristics of the refrigerant. For a given heat pump the COP varies with the temperature of the input and the output. Typically, the COP is quoted at specified input and output conditions – e.g. B0/W35 means an input water temperature to the evaporator of 0°C, and an output water temperature from the condenser of 35°C.

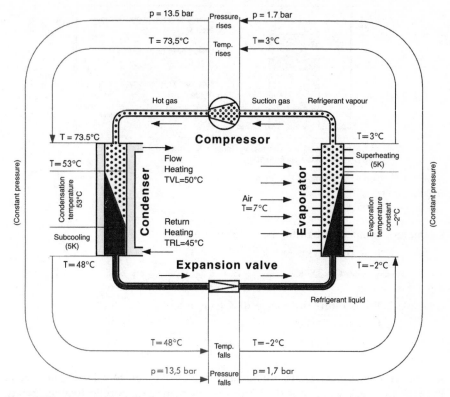

Figure 2.3 Heat pump refrigeration cycle
Source: Ochsner

2.4 Carnot Cycle

The heat pump cycle more or less reverses the (ideal) Carnot Cycle for a combustion engine. This means that the COP can also be calculated using the temperature difference between the heat source (evaporator) and the heat sink (condenser):

$\varepsilon_c = T/T - T_u = T/\Delta T$

ε_c = COP from Carnot (Carnot efficiency)

T_u = Temperature of environment out of which the heat is to be collected (specifically, T_o evaporator temperature)

T = Temperature of environment into which the heat is distributed (specifically, condenser temperature)

ΔT = Temperature difference between the warm and cool sides

(all temperatures given in absolute temperature, in degrees Kelvin, or °K)

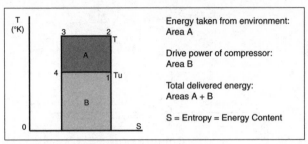

4-1 Evaporation / 1-2 Compression (Temperature Increase)
2-3 Condensation / 3-4 Expansion

Figure 2.4 The Carnot Cycle in the T-S diagram
Source: Ochsner

Example 1: Temperature difference 50 K (simplified)

$$T_u = 0°C = 273 \text{ K}$$

$$T = 50°C = 323 \text{ K} \qquad \varepsilon_c = \frac{T}{T-T_u} = \frac{323}{323-273} = 6.46$$

Example 2: Temperature difference 30 K (simplified)

$$T_u = 0°C = 273 \text{ K}$$

$$T = 30°C = 303 \text{ K} \qquad \varepsilon_c = \frac{T}{T-T_u} = \frac{303}{303-273} = 10.1$$

In practice, ideal processes are not possible. The COP for an actual heat pump cycle includes various losses and is therefore smaller than the theoretical value. Due to thermal, mechanical and electrical losses, as well as the power demand of the auxiliary pump, the achieved COP is smaller than the Carnot Efficiency. As an approximation, the actual COP can be taken as 0.5 × the Carnot Efficiency.

Example 1:	$\varepsilon_c = 6.46$	$\varepsilon = 3.23$
Example 2:	$\varepsilon_c = 10.1$	$\varepsilon = 5.05$

TEMPERATURE DIFFERENCE DETERMINES THE COP

In all cases, the Coefficient Of Performance depends on the temperature difference between the heat source and the heat sink (see Figure 2.5). The smaller the required temperature difference, the more efficiently and economically the heat pump operates – because the compressor has to do less work to lift the temperature of the refrigerant gas. Therefore, optimization of the entire system is extremely important.

The COP is also dependent on other factors such as the temperature differences within the heat collection and distribution systems. These effects should be taken into account when comparing manufacturer and test data.

SEASONAL PERFORMANCE FACTOR (ANNUAL EFFICIENCY)

The annual efficiency indicates the total amount of heating energy delivered in an annual heating period in relation to the total electrical power consumed in the same

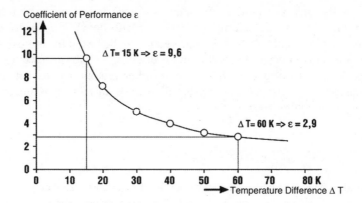

Figure 2.5 Coefficient of performance with respect to temperature difference
Soure: BWP

period. Furthermore, working efficiencies can be defined for the heat pump alone or for the complete heating system. The average COP over a heating (or cooling) season is often referred to as the Seasonal Performance Factor (SPF). See VDI 4650.

Calculation of
- Seasonal Efficiency Ratios (SEER)
- Heating Seasonal Performance Factor (HSPF)

In North America
- see ANSI/ASHRAE Standard 116–1995 (RA 2005) – see Figure 15.20

2.5 Working fluid/refrigerant

For working fluid (refrigerant), suitable substances are those with large specific heat capacities and which evaporate at low temperatures.

Today, only chlorine-free refrigerants are permitted. These are non-ozone depleting refrigerants (Ozone Depletion Potential, ODP = 0). R 134a, R 407C, R410A, R404A and propane fulfil these conditions. In domestic applications, R 134a, R 407C and other blends are often used as they are both non-flammable and non-toxic. To provide compressor lubrication, biodegradable esther oil is used in modern heat pumps, further minimizing any potential environmental impact. Even in systems with direct vaporization, the theoretical possibility of environmental damage is practically eliminated.

Heat pumps with inflammable refrigerants (e.g. propane) are subject to numerous safety guidelines and are therefore restricted with respect to installation location.

Systems with CO_2 as refrigerant are still in development, mainly due to the high pressure which this technology requires.

Most of the heat pumps that are being discussed here have their refrigerant installed during the manufacturing process – as with domestic refrigerators. It is therefore extremely unlikely that the refrigerant will be lost to the atmosphere – as long as care is taken when the heat pump is disposed of. However, concerns over the large losses of refrigerant that occur in other sectors means that there are increasing requirements on heat pump installers to be fully aware of, and trained in, controlling potential refrigerant losses.

2.6 Enthalpy-pressure diagram

The enthalpy, h, is a measure of energy contained in a substance. The progression of the ideal cycle follows the path 1 2 2′ 3 4. The vapour dome shows the separation between the liquid phase (left), unsaturated-vapour (middle) and super-heated vapour phases (right) (Figure 2.6).

In an ideal cycle, the refrigerant behaves as an ideal gas and all processes occur without losses (isentropic).

In *real cycles*, the compression does not occur along the line 1–2 (isentropic), but due to losses to a somewhat higher compression temperature at the same saturation pressure. Therefore, more compression work is required in order to achieve the same end pressure and saturation temperature. The energy transferred in the cycle can be taken directly as the enthalpy differences from the h, lg p-diagram (Figure 2.6). The Carnot Efficiency can be quickly determined using these values:

$\varepsilon_c = h2 - h3/h2 - h1$

For actual processes, the COP may be determined as:

$\varepsilon_c = h2^* - h3^*/h2^* - h1^*$

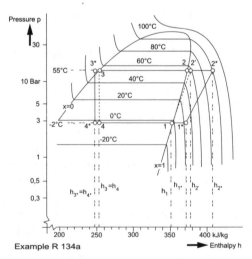

Figure 2.6 Determination of the COP in the h, lg p-diagram
Source: Ochsner

Cycle with super-heating and sub-cooling:

4* – 1   Evaporation, absorption of vaporization energy h1 – h4
1 – 1*   Super-heating of intake gas
1* – 2*   Compression to set compression temperature (super-heated refrigerant vapour)
2* – 2   Cooling to saturated vapour temperature, release of super-heating energy h2* – h2
2 – 3   Condensation, release of vaporization energy, h2′ – h3
3 – 3*   Sub-cooling of fluid
3* – 4*   Expansion in the unsaturated vapor phase; no energy release (transformation from sensible to latent heat)

Cycle without super-heating and sub-cooling:

4 – 1 – 2′ – 3 – 4

2.7 Heat pump cycle with injection cooling

In order to increase COP and heating capacity and to make a higher temperature-lift possible, heat pumps can be designed with injection cooling, e.g. air source heat pumps with vapour-injection – cooling can supply temperatures up to 65°C even in coldest climates.

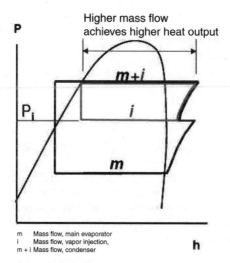

m   Mass flow, main evaporator
i   Mass flow, vapor injection,
m + i Mass flow, condenser

Figure 2.7 Refrigerant Cycle with Enhanced Vapour Injection
Source: Copeland GmbH

3

Heat Pump Types

Heat pumps can be categorized according to:

■ function:
heating, cooling, domestic water heating, ventilation, drying, heat recovery etc.

■ heat source:
ground, ground water, air, exhaust air etc.

■ working fluids – heat source/heat distribution:
brine/water, water/water, direct-expansion/water, air/water, air/air etc.

■ unit construction:
compact, split
installation location (indoor, outdoor)
compression heat pump
absorption heat pump
drive power (electric, gas)
number of compression stages
etc.

Air conditioners and refrigerators are also heat pumps, making use of the 'cold side' of the heat pump cycle.

The following is a short summary of the standard electrically driven, vapour compression type heat pumps which are available for applications in single and multiple family homes, and in industrial, commercial and community buildings. Due to the diverse range of manufacturers' construction methods, this should only be considered as an overview.

3.1 Brine/water, water/water heat pump

APPLICATIONS
Both brine/water and water/water heat pumps are used for monovalent heating operation (usually ground source), as well as cooling, heat recovery and domestic

hot water production. Water/water (WW) heat pumps are used with an available water source, e.g. aquifer-fed borehole, lake, or water body. Brine/water (BW) heat pumps are used in closed loop ground coupled installations. The 'brine' refers to the fact that these systems can run at low source temperatures, often below freezing – so it is necessary to have an antifreeze solution in the source water circuit. This was traditionally a brine solution, but today is usually a water/antifreeze mixture.

CONSTRUCTION/HOUSING
Brine/water and water/water heat pumps are generally compact units for indoor installation, and consist of the following components (see Figures 3.2 and 3.3).

3.1.1 Refrigeration cycle
Generally a fully hermetic compressor (piston or scroll for extremely quiet operation) with built-in, internal overload protection is used in these heat pumps. Stainless steel flat plate heat exchangers or other types like shell and tube or coaxial are used for the evaporator and condenser.

Other items in the refrigerant cycle are the expansion valve and possibly a sight glass, accumulator and a filter/drier. The refrigeration cycle (Figure 3.1) should be fully insulated against thermal losses and prevent condensation inside the heat pump. For heating/cooling reversible operation, see Figure 10.3.

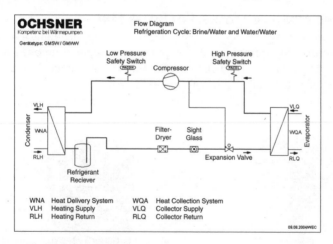

Figure 3.1 Refrigeration cycle of a brine/water or water/water heat pump
Source: Ochsner

3.1.2 Refrigerant
Chlorine-free working fluids which do not cause damage to the ozone layer and have low global warming potential are to be used. With non-flammable refrigerants (R 407C, R 404A, R 410A, R 134a etc.), installation in any location is possible. Inflammable refrigerants (R 290 and others) are uncommon due to safety requirements.

The non-flammable safety refrigerants also enable use of biodegradable, synthetic refrigerant oil (esther oil).

3.1.3 Electrical components and controller

Electrical components and the controller are either integrated or externally mounted, depending on manufacturer and model. Control of the heating system is commonly integrated. A micro-processor driven heating controller should be weather-and-load-dependent. The heating curve as well as many functions and programmes can be individually set.

3.1.4 Safety measures

Motor protection relays, high and low refrigerant pressure switches, and a frost protection thermostat are used.

3.1.5 Display

These vary considerably with heat pump manufacturers depending on the simplicity or complexity of the controller. Typically, any or all of the following will be found: Temperature display for heat pump supply and heat source supply, illuminated operation switch, operation-hour counter, warning indicators for high-pressure, low-pressure, frost-protection, motor-relays and more, and reset button.

HEATING AND BRINE ACCESSORIES, HOT WATER TANK

In numerous models, the heating system circulation pump (including a check valve) and/or the source side (brine) circulation pump are integrated into the heat pump casing to simplify installation and minimize the space requirements. In some compact units the domestic hot water tank is also built in.

(a)   (b)

Figure 3.2 Water/water heat pump Golf series (midi): (a) closed and (b) open unit
Source: Ochsner

Figure 3.3 Brine/water heat pump
Source: AWP

Figure 3.4 Brine/water heat pump with integrated hot water tank,
Source: Ochsner

In modern domestic heat pumps attention will have been paid to minimizing noise and vibration arising from the compressor operation. This will take the form of acoustic shielding in the casing, together with vibration isolation of the compressor from the casing. The use of flexible hydraulic connections and anti-vibration feet further reduce any noise and vibration.

3.2 Direct expansion/water heat pump

Direct expansion heat pumps do not have an intermediate heat exchanger on their source side. Instead, a loop of suitable pipe containing the refrigerant and lubricant is put in direct contact with the ground or water body. The compressor operation circulates the refrigerant directly around this loop – thus eliminating the heat transfer losses associated with the intermediate water/DX heat exchanger found in conventional water source heat pumps. There is also no need for a source side circulation pump – the compressor undertakes this role. However, care has to be

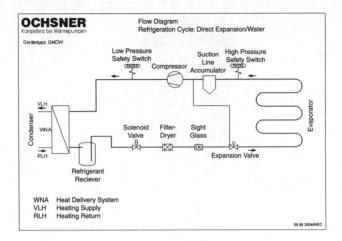

Figure 3.5 Heat Pump cycle of Direct Expansion system (simplified)
Source: Ochsner

taken to ensure that the DX loops are totally sealed, corrosion resistant, and that the lubricant is adequately circulated to meet the needs of the compressor.

Environmental authorities in different countries are enforcing special standards on these systems depending on local concerns over increasing volumes of refrigerant being installed in vulnerable ground loop arrays. For example, there are double-walled continuous ground loops and solenoid valves that are automatically closed when the compressor is not operating.

APPLICATIONS
Direct expansion/water heat pumps are used for monovalent heating operation (with ground source) and for domestic hot water preparation.

CONSTRUCTION/HOUSING
Installation of these compact units is unrestricted indoors if non-flammable refrigerants are used (see DIN 8901).

3.2.1 Refrigeration cycle
Generally, a fully-hermetic compressor (piston or scroll for extremely quiet operation) with built-in, internal over-load protection. Usually stainless steel flat plate heat exchanger for condenser. Collector tubing for evaporator loop. Expansion valve. The refrigeration cycle should be fully insulated against thermal losses and prevent condensation. The refrigerant receiver should be generously dimensioned (see Figure 3.5).

3.2.2 Refrigerant
Chlorine-free working fluids which do not cause damage to the ozone layer and with low global warming potential are to be used. With non-flammable

refrigerants (R 407C, R 404A, R 410A, R 134a etc.), installation in any location is possible.

The non-flammable safety refrigerants also enable use of biodegradable, synthetic refrigerant oil (esther oil).

Use of inflammable refrigerants (R 290, and others) is restricted to outdoor installation. For combination with CO_2 loops, see Chapter 14.8.

3.2.3 Electrical components and controller

Electrical components and the controller are either integrated or externally mounted, depending on manufacturer and model. Control of the heating system is commonly integrated. A micro-processor driven heating controller should be weather and load dependent. The heating curve as well as many functions and programmes can be individually set.

3.2.4 Safety measures

Motor protection relays, high and low pressure switches. A solenoid valve should be included which in the event of collector damage automatically closes to prevent refrigerant in the heat pump from draining into the ground.

3.2.5 Display

These vary considerably with heat pump manufacturers depending on the simplicity or complexity of the controller. Typically, any or all of the following will be found: Temperature display for heat pump supply and heat source supply, illuminated operation switch, operation-hour counter, warning indicators for high-pressure, low-pressure, frost-protection, motor-relays and more, and reset button.

HEATING ACCESSORIES

In numerous models, the heating system circulation pump (including check valve) is integrated into the heat pump and is operation ready.

Figure 3.6 shows a direct expansion/water heat pump and the separate multi-function unit Europa for domestic water heating and ventilation.

3.3 Direct expansion/direct condensation heat pump

In this variation of the direct expansion system, the distribution of heat is achieved directly using the refrigerant, e.g. in a radiant floor heating system. The condenser, like the evaporator, is composed of seamless, plastic-sheathed copper tubing. The latent condensation energy is released at a constant temperature.

Instead of individual room control, zone control of the individual loops only is possible. Careful attention must be given to the sizing and quality of copper tubing, as well as the other components. Obviously, considerable care has to be taken in carrying out the load side installation in order to prevent loss of refrigerant due to leaks or damage.

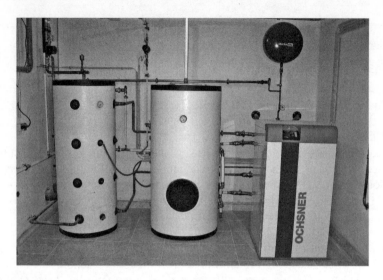

Figure 3.6 Heat Pump – Direct Expansion/Water with Buffer Tank and Hot Water Tank
Source: Ochsner

3.4 Air/water heat pump – split units

APPLICATIONS
Air/water heat pumps use outside air as their heat source and are mostly operated in bivalent heating systems, as well as for cooling, heat recovery and domestic hot water preparation. They are particularly well suited for retrofit and renovation applications (see Chapter 14.1).

CONSTRUCTION/HOUSING
The indoor unit contains the substantial components and is fitted indoors, protected from weather and freezing temperatures. The outdoor unit is connected to the indoor unit via refrigeration lines. Through the elimination of air ducts, extremely quiet, energy efficient fans are made possible (see Figure 3.7).

3.4.1 Refrigeration components – indoor unit
Generally a fully-hermetic compressor (piston or scroll for extremely quiet operation) with built-in, internal overload protection is used in these heat pumps. Stainless steel flat plate heat exchangers are used for the condenser. Expansion valve, weather-dependent defrost mechanism – preferably hot gas. The refrigeration cycle should be fully insulated against thermal losses and prevent condensation.

Designs with vapour-injection cooling are suitable for use to 65°C flow temperature.

3.4.2 Refrigeration components – outdoor unit
Copper-tube aluminium finned evaporator. For quiet operation an axial fan with low

Figure 3.7 Air/Water Heat Pump (Split-Configuration)
Source: Ochsner

speed should be used. Split systems allow for the use of large, high capacity evaporators and quiet, slow rotating, energy efficient fans.

Follow manufacturer guidelines for the connection of the units.

3.4.3 Refrigerant

Chlorine-free working fluids which do not cause damage to the ozone layer and with low global warming potential are to be used. With non-flammable refrigerants (R 407C, R 404A, R 410A, R 134a etc.), installation in any location is possible. Inflammable refrigerants (R 290 and others) are uncommon due to safety requirements.

The non-flammable safety refrigerants also enable use of biodegradable, synthetic refrigerant oil (esther oil).

3.4.4 Electrical components and controller

Electrical components and the controller are either integrated or externally mounted, depending on manufacturer and model. Control of the heating system is commonly integrated. A micro-processor driven heating controller should be weather and load dependent. The heating curve as well as many functions and programmes can be individually set. An electronic controller for the defrost mechanism is generally provided. Demand activated defrosting increases system performance.

3.4.5 Safety measures

Motor protection relays, high and low refrigerant pressure switches, and thermostat for defrost control.

3.4.6 Display

These vary considerably with heat pump manufacturers, depending on the simplicity or complexity of the controller. Typically, any or all of the following will be found:

Temperature display for heat pump supply, illuminated operation switch, operation-hour counter, warning indicators for high-pressure, low-pressure, motor-relays and more, and reset button.

3.5 Air/water heat pump – compact units, indoor installation

APPLICATIONS

Air/water heat pumps use outside air as their heat source and are mostly operated in bivalent heating systems, as well as for cooling, heat recovery and domestic hot water preparation.

CONSTRUCTION/HOUSING

Compact unit suitable for indoor installation (see Figure 3.8).

3.5.1 Refrigeration cycle

Generally a fully-hermetic compressor (piston or scroll for extremely quiet operation) with built-in, internal over-load protection. Copper-tube aluminium finned evaporator, axial fan for silent operation and short ducts, otherwise radial fan, defrost mechanism (preferably hot-gas), drain pan (preferably heated), stainless steel plate heat exchanger for condenser, expansion valve. The refrigeration cycle should be fully insulated against thermal losses and prevent condensation.

Designs with vapour-injection cooling are suitable for up to 65°C flow temperature.

3.5.2 Refrigerant

Chlorine-free working fluids which do not cause damage to the ozone layer and with low global warming potential are to be used. With non-flammable refrigerants (R

Figure 3.8 Air/water heat pump for indoor installation
Source: AWP

407C, R 404A, R 410A, R 134a etc.), installation in any location is possible. Inflammable refrigerants (R 290 and others) are uncommon due to safety requirements.

3.5.3 Electrical components and controller

Electrical components and the controller are either integrated or externally mounted, depending on manufacturer and model. Control of the heating system is commonly integrated. A micro-processor driven heating controller should be weather and load dependent. The heating curve as well as many functions and programmes can be individually set. An appropriate control device for the defrost function is necessary. Demand activated defrosting increases system performance.

3.5.4 Safety measures

Motor protection relays, high and low refrigerant pressure switches, and thermostat for defrost control.

3.5.5 Display

These vary considerably with heat pump manufacturers depending on the simplicity or complexity of the controller. Typically, any or all of the following will be found: Temperature display for heat pump supply, illuminated operation switch, operation-hour counter, warning indicators for high-pressure, low-pressure, motor-relays and more, and reset button.

3.6 Air/water heat pump – compact units, outdoor installation

Figure 3.9 shows that compact outdoor units can be installed in any open location.

Figure 3.9 Air/water heat pump – outdoor setup
Source: Ochsner

All components of the refrigeration cycle are integrated into the unit, so no air ducts are required. The housing must provide protection against weather elements.

The connection to the heating system in the building consists of two insulated pipes for the supply and return. These are to be installed underground with the heat pump power supply and control cable. Noise emissions are dependent upon sizing, construction and speed of the fan. Large, slow rotating fans operate quietly, insulation panels further reduce noise. Follow manufacturer instructions regarding orientation. The defrost function for outdoor temperatures to approximately –16°C is usually achieved with hot gas reverse cycle.

Designs with vapour-injection cooling are suitable for up to 65°C flow temperature.

3.7 Domestic hot water/heat pumps – air source, compact units

CONSTRUCTION
These are supplied as fully integrated compact units, normally as exhaust air heat pumps with a 200–300 litre hot water storage tank. All components such as the compressor, condenser, evaporator and controller are integrated into the unit (see Figures 3.10–3.13).

3.7.1 Refrigeration cycle
Full-hermetic compressor, finned evaporator, axial or radial fan. The air intakes and exhaust can be placed in different locations. Air ducts up to 20 metres in length are acceptable (for radial fan). Condenser located in tank, in heat pump together with

Figure 3.10 Multi-Function Unit Europa 312 with Tiptronik E (temperature display, illuminated operation switch). With the Tiptronik, all temperatures, operating conditions and parameters are displayed on an LCD display. A hot-gas defrost mechanism enables operation at any outdoor temperature. Ventilation is independent of heat pump compressor operation
Source: Ochsner

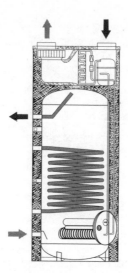

Figure 3.11 Section: Domestic Hot Water Heat Pump with Condenser-Coil and Heat Exchanger for optional Heat source
Source: Ochsner

feed pump or external. Optional defrost mechanism (hot-gas) for operation at all outdoor temperatures.

3.7.2 Refrigerant

Usually, environmentally friendly, chlorine-free and non-flammable safety refrigerant R 134a with biodegradable esther oil is used.

Figure 3.12 Air/Water Heat Pump – Compact Unit AWP/ALKO
Source: AWP

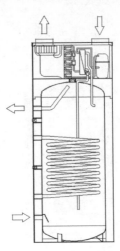

Figure 3.13 Section: Domestic Hot Water Heat Pump with Loading-Pump and Heat Exchanger for optional Heat source
Source: AWP

3.7.3 Electrical components

The most sophisticated units offer the following functions: fully automatic defrost control, programmable heat-up function, auxiliary electric element, Legionella function for thermal water treatment on demand, use of off-peak power, independent fan function for ventilation, diagnostic display, and many more.

Auxiliary heating with external boiler:
For configurations with tanks with built-in heating coils, alternative water heating with auxiliary separate boiler or solar pannel is possible.

Auxiliary electrical element:
Many units have an electrical element (wet or dry) for elevated hot water demand and back up.

3.8 Domestic hot water heat pump – air source, split units

With split domestic hot water units, external domestic hot water storage tanks are used. A built-in feed pump will circulate water from the heat pump through the hot water tank. In this way, existing tanks of any model or capacity can be used (advantage: desired storage volume).

Heat pumps with ca. 2kW capacity can be usefully combined with tanks from 200 to 500 litre volume. Units with ca. 5kW capacity can be used for commercial and industrial applications with 1000 to 2000 litre storage volume. There are split heat pumps which stand next to the tank and others are mounted directly on to the tank (see Figures 3.14 and 3.15).

Figure 3.14 Split unit
Source: Ochsner

Figure 3.15 Insert heat pump
Source: Thermo Energie

Figure 3.16 Split unit with internal flat plate heat exchanger
Source: Ochsner

Additionally, split units with internal flat plate heat exchangers are available (Figure 3.16).

3.9 Ground/water heat pump – domestic hot water heat pump – ground source, split units

Ground source split units use heat stored in the ground (Figure 3.17).

Figure 3.17 Module mini + ground loop
Source: Ochsner

3.10 Air/air heat pump – ventilation

Heat pumps are also found in controlled dwelling ventilation applications. This enables an increase in heat recovery from the exhaust air and can even allow for cooling of selected rooms. For these applications, various units are used. The air/air heat pumps have a full-hermetic compressor, finned heat exchanger for evaporator and condenser, as well as an expansion valve and necessary safety mechanisms.

The control of the unit is achieved with a simple room and supply air thermostat or complex electronics. Figures 3.18 and 3.19 show ventilation heat pumps from different firms.

3.11 Exhaust-air heat pumps, additional designs

There are various other designs on the market which use heat from exhaust air for heating and for water production.

Hydronic systems as well as air ventilation is used for heat distribution.

Direct electric heating is normally used as an additional heat source, but this may result in increased heating costs.

Figure 3.18 Controlled dwelling ventilation with kwl-centre
Source: Helios

Figure 3.19 Controlled dwelling ventilation, auxiliary heating and cooling – Europa 122
Source: Ochsner

3.12 Heat pumps for air heating/cooling

Air-based heating systems are common in America and Southern Europe. In these systems the heat distribution via the heat pump occurs using air ducts, not using a hydraulic system (see Figure 3.20).

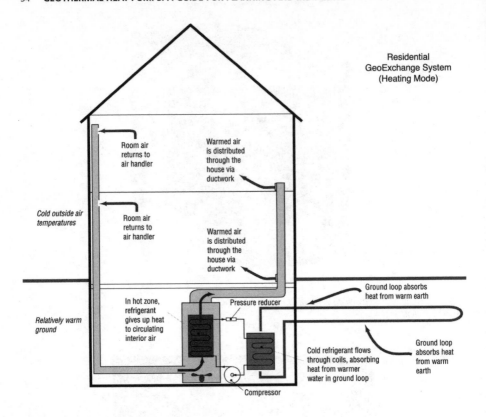

Figure 3.20 Residential GeoExchange System
Source: GeoExchange

4

Complete System Planning

4.1 Planning a heat pump heating system

Figure 4.1 depicts a heat pump heating system, consisting of a heat source (HS), a heat pump (HP) and a heat delivery system (HD).

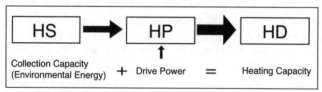

Collection Capacity
(Environmental Energy) **+** Drive Power **=** Heating Capacity

Figure 4.1 Heat pump system – heating function
Source: DIN

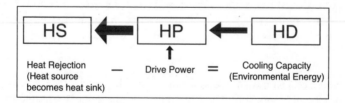

Heat Rejection
(Heat source **—** Drive Power **=** Cooling Capacity
becomes heat sink) (Environmental Energy)

Figure 4.2 Heat pump system – cooling function   Cooling function (see Chapter 10)
Source: DIN

For system planning all components must be designed to interact optimally, in order to ensure the highest level of performance and reliable operation. Critical design data should be supplied in manufacturer's data sheets, and the system should adhere to quality control standards and installation recommendations also found in the manufacturer's technical guide.

Terminology: For instance water/water means heat source water and hydronic heat delivery.

4.2 Heat source selection

As a rule, the heat source with the highest temperature levels should be selected (Figure 4.3). This ensures the highest possible coefficient of performance and thereby the lowest operation costs.

WATER
If groundwater is available at a reasonable depth and temperature, at acceptable quality and in sufficient quantity, the highest COP can be achieved. The best groundwater heat source system is an open system and may require approval. Regulations concerning the use of groundwater resources differ from country to country and region to region. Attention must be paid to these local require-ments and regulations before an open loop groundwater system is installed and operated.

If groundwater is not available, the ground heat source may be an option.

GROUND (BRINE)
If the use of groundwater is not available, the ground can function as an efficient, effective thermal storage medium with a relatively high temperature level. If a sufficient surface area is available, horizontal collectors offer the most cost effective solution (approval may be required). If space is limited, vertical loops using geothermal energy are an effective solution. These heat sources are closed systems, meaning that the brine (i.e. antifreeze solution) stays within the buried tube system. Heating and cooling options are possible.

GROUND (DIRECT EXPANSION)
In direct expansion systems, the warmth stored in the ground is absorbed directly by the working fluid (refrigerant). This results in an increased coefficient of performance. Horizontal collectors are mainly used with this system.

AIR
If groundwater or ground source systems cannot be used, air as a heat source is available practically anywhere. These systems are particularly suitable for retrofits or in combined operation with another heat source (i.e. bivalent operation). Heating and cooling options are possible. (For details on all heat sources, see Heat Source Planning Tips.)

4.3 Heating system selection

In this book hydronic heat distribution systems are described.

Determination of the maximum heating supply temperature
As a rule, the lower the heating system temperature, the higher the coefficient of performance of the heat pump and the lower the heating costs. In order to achieve

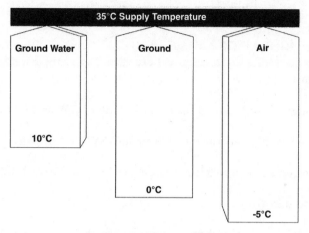

Figure 4.3 Example temperature lift for groundwater, ground and air-source heat pumps
Source: BWP

this, the largest possible surface area for heat transfer must be selected. Ideal for this criteria are low temperature radiant floor and wall heating systems (i.e. max 35°C supply temperature).

Furthermore, low temperature radiant heat ensures maximum comfort.

For new buildings, for both economic and comfort reasons, a radiant surface heating system should be selected. These offer the additional benefit of effective use of thermal mass during possible electric cut-off times and to make use of low tariff periods.

For old buildings, low temperature radiators are generally selected. These are laid out for a maximum supply temperature of 55°C to 65°C (see manufacturer data). A combination of both systems is also possible.

4.4 Heat pump selection

The selection of the heat source determines to a great extent the required heat pump type and operation. In order to determine the required heating capacity of the heat pump, the following procedure should be followed.

4.4.1 Determination of heating demand
With heat pumps, exact size selection is important because oversized systems operate with lower efficiencies, leading to excessive costs.

The determination of the heating demand is to be completed according to relevant national standards, e.g.
Germany: DIN 4701, EnEV 2002
Austria: ÖNORM 7500 (B 8135)
Switzerland: SIA 380-1 (2001 Ed.), SIA 384.2

United Kingdom: National Calculation Method (NCM), or Standard Assessment Procedure (SAP)

United States: ASHRAE GRP 158 (updated)

The following values are based upon experience in European homes (heating demand, W/m²):

■ Old building with standard (at the time) insulation 75 W/m² to 100 w/m²

■ New building with good insulation (German WSVO 95) 50 W/m²

■ Low energy house, new building according to German EnEV 40 W/m²

■ Passive house 15 W/m².

The specific heating demand (W/m²) is multiplied by the area of heated living space. The result is the total heating demand, which accounts not only for the transmission but also for the ventilation heating demands (see chapter on dwelling ventilation).

Note: The annual heating demand in kWh per year indicates the heating energy to be delivered over the course of a year. This must be taken into account when sizing the heat collector for a closed loop ground coupled system and during initial operation (drying the residual moisture in building materials with a ground source system should be avoided). The building should stand for one winter in order to fully dry the building materials. Otherwise, the heating demand could be as much as 50 per cent higher. Remedy: not heating certain areas (i.e. basement), drying building materials or placing an auxiliary electrical heating element in a buffer storage tank.

If a heat pump system with the **cooling option** is selected, additional information is found in Chapter 10. In this operation configuration, the heat source becomes a heat sink and vice versa.

4.4.2 Utility interruptible rates

Utilities sometimes offer a reduced price for electricity for a heat pump. In return, they maintain the right to interrupt the supply at certain periods during the day. The power supply can be interrupted, for example, 3 × 2 hours within a 24-hour period. Therefore, the heat pump must meet the required daily heat demand in the time in which power is delivered.

Example:

Theoretical planning for 3 x 2 hour cut-off: Calculated heating demand without cut-off: 9 kW
                                                 Max. cut-off time: 3 x 2 = 6 hours
                                                 Operation time = 18 hours

Theoretical heating demand with cut-off:

$$\frac{9\ kW \times 24\ h}{18\ h} = 12\ kW$$ (This represents a 33 per cent increase in heating demand.)

Practical tip: the calculation of the heat requirement includes fully heating all rooms during the lowest possible outdoor temperature, a situation which rarely occurs.

Experience has shown that a 20 per cent increase is sufficient, particularly for radiant floor heating systems.

9 kW • 1.2 = 10.8 kW

In every case, local conditions are to be considered.

4.4.3 Domestic hot water heating

The documentation in section 12.1.6 should be adhered to for domestic hot water heating with a space heating heat pump (0.25 kW per person = average requirement for a single family home).

Once the total required heating capacity of the heat pump is determined, an appropriate heat pump model can be selected, based upon the technical performance data provided by heat pump manufacturers.

The operation configuration is to be taken into consideration as well.

4.4.4 Operation configurations

The following operation configurations are possible:
– **Monovalent** (heat pump without any auxiliary heat source)
– **Bivalent-Parallel** (heat pump + auxiliary heating)
(The auxiliary heating is usually with an electrical heating element, referred to as mono-energetic.)
– **Bivalent-alternate** (heat pump or auxiliary heating)

MONOVALENT (see Figure 4.4)

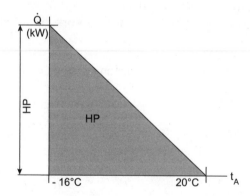

Figure 4.4 Monovalent operation
Source: Ochsner

The heat pump is the only heat source, providing 100 per cent of the heating demand at all times. Well suited for applications with supply temperatures up to 65°C. Systems with groundwater or ground heat source collectors are operated as monovalent systems.

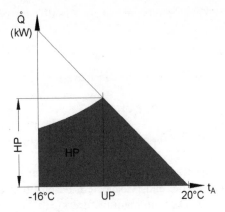

Figure 4.5 Bivalent-parallel operation
Source: Ochsner

BIVALENT-PARALLEL (see Figure 4.5)

The heat pump heats independently to a certain set point, at which an auxiliary heating system (electric element or boiler) is turned on and the two systems operate in parallel to meet heating demand. For maximum supply temperature up to 65°C. Used mainly with new air source systems or renovation of old buildings.

BIVALENT-ALTERNATE (see Figure 4.6)

The heat pump heats independently to a certain set point. Once this point is reached, a boiler meets the full heating demand. Suitable for supply temperatures up to 90°C. Normally installed in the case of renovation.

Figure 4.7 depicts the portion of annual heating work for a heat pump in bivalent operation.

Example of Figure 4.7: For a start-up temperature of 0°C, the air/water heat pump produces 88 per cent of the annual heating demand even at outer temperatures as low as –16°C.

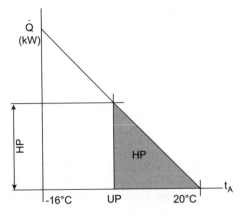

Figure 4.6 Bivalent-alternate operation
Source: Ochsner

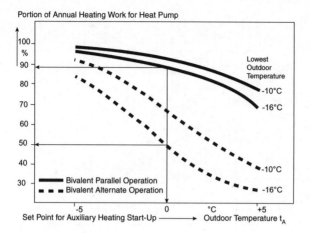

Portion of Annual Heating Work for Heat Pump

Figure 4.7 Portion for bivalent operation
Source: BWP

The lower the external start-up temperature for the auxiliary system, the higher the percentage of annual heating demand produced by the heat pump. The annual heating demand is also dependent on the climate zone and operation configuration.

4.4.5 Heat pump selection

4.4.5.1 SELECTION FOR MONOVALENT OPERATION
The selected heat pump model must have the capacity to cover the total heating demand at the given heating supply temperature. Care must also be taken to select the heat pump at the appropriate source side temperature; i.e. the worst day entering water temperature for a water/water (WW) system may be 8°C, but a closed loop (BW) system at the same location may have an EWT of only –2°C, say.

Test conditions according to the ANSI Standard (see Figure 15.19, page 139).

4.4.5.2 SELECTION FOR BIVALENT-PARALLEL OPERATION
The portion of the total heating demand to be covered by the heat pump at the given heating supply temperature must be determined (see Figure 4.8).
The results of the graphical determination (Figure 4.8):

■ Heating capacity of the heat pump at the design point is 5.5 kW

■ Min. required heating capacity of the electric heating element is 6.5 kW

■ Start-up point of second heat source, –1°C.

Rule of thumb: At a start-up temperature of 0°C, the electric heating element should have a heating capacity approximately equal to that of the heat pump (at this operating point) for climates with –10°C or lower temperatures.

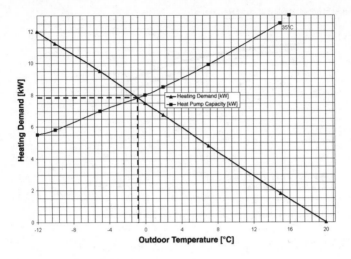

Figure 4.8 Determination of the required electrical element heating capacity and the start-up point for bivalent-parallel operation
Source: Ochsner

4.4.5.3 SELECTION FOR BIVALENT-ALTERNATE OPERATION

For this operation configuration, the maximum supply temperature of the heat pump usually serves as the change-over point (see Figure 4.9).

Once the required heating capacity at the change-over point is known (Figure 4.10), a suitable model can be selected using the heat pump performance curves provided. The outdoor temperature at the change-over point and the corresponding heating supply temperature are critical factors in this determination.

4.5 Retrofit/renovation

The heating demand of existing buildings can be determined from the annual oil or gas consumption using the following rules of thumb:

$\dot{Q}(kW)$ = Oil consumption (l/a)/250 (l/a kW)
$\dot{Q}(kW)$ = Gas consumption (m³/a)/250 (m³/a kW)

The factor of 250 results from the heating capacity (10kWh/l), annual operation-hours (1900), and the annual system efficiency (0.75).

In combination with building renovation, the implementation of a heat pump is also feasible for older buildings. Improved insulation reduces the specific heating demand and allows for lower heating supply temperatures. Moreover, many existing heating systems, especially the radiators and piping network, are over-sized. These existing systems may be suitable for the operation conditions required with a heat pump.

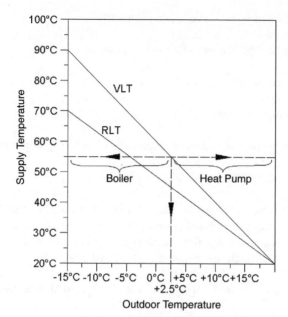

Example: Heating system with supply temp. 90°C and
return temp. 70°C, heat pump max. supply temp. 55°C,
results in change-over temperature of +2.5°C

Figure 4.9 Determination of the change-over point for bivalent-alternate systems using supply temperature
Source: Ochsner

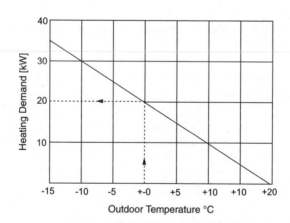

Example: Determination of heating demand at a change-
over temperature 0°C for a heating demand of
35kW at -15°C for sizing of heat pump.

Figure 4.10 Determination of heating demand at the change-over point
Source: Ochsner

The following data must be surveyed or calculated:

■ The heating demand before and after new insulation.

■ Previous maximum supply temperature required.

■ Heating system piping diameter.

The following must be considered when implementing a heat pump system:

■ Air source heat pumps can be installed anywhere.

■ Electric heating elements for mono-energetic bivalent-parallel operation have low investment costs and can provide additional supplementary heating. However, care must be taken to ensure that heating costs and carbon emissions do not become excessive through the use of this additional direct electric heating.

■ Heat pumps with a safety refrigerant can be installed anywhere without restrictions.

■ For heating with radiators or zoned underfloor or air source, a buffer storage tank should be installed.

■ The operating limit for state of the art systems with R 407C and for R 134a is max. 65°C.

■ Check for feasibility of bivalent-alternate operation (in situations where the old boiler still functions properly).

■ The availability of suitably sized electrical supplies in situations where there is only a single phase supply.

Nominal Conditions for performance test according to DIN, ÖN, EN:

Heat source system   /   Heating system

Water/Water
Heat source water
Heat source temperature 10°C

W 10   /   W 35

Heat use: Water
Flow temperature 35°C

Brine/Water
Heat source brine
Heat source temperature 0°C

B 0   /   W 35

Heat use: Water
Flow temperature 35°C

Ground/Water
Heat source temperature 4°C
(complies with brine = 0°C)

E 4   /   W 35

Heat use: Water
Flow temperature 35°C

Air/Water
Heat source air
Heat source temperature 2°C

A 2   /   W 35

Heat use: Water
Flow temperature 35°C

5

Planning Instructions for Ground Heat Source – Brine Systems (Horizontal Collector, Trench, Vertical Loop)

5.1 Ground heat source

The ground serves as an ideal heat source for monovalent heat pump systems. The ground stores solar energy and is also regenerated with rain water. As a result, sufficient energy is available in winter, even when the ground is covered in snow. Vertical loops utilize more storage and a significant portion of geothermal energy. Temperatures at 15m underground remain constant the whole year, being similar to the average ambient heat air temperature over the year. They are, for instance, for Scandinavia 2 to 9°C, for Germany 9 to 11°C and for Italy 13 to 17°C.

5.1.1 System description (see Figure 5.1)

The heat source (or heat sink) is a closed system consisting of a horizontal ground collector or a vertical loop. The heat transfer medium, brine (water with anti-freeze), is circulated through the collector or loop and the heat pump with a brine pump.

Plastic piping with suitable wall thickness to resist damage is selected. The method of installation varies and must adhere to any applicable standards. In some cases, permits must be obtained.

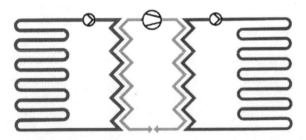

Figure 5.1 Simplified operating scheme brine/water
Source: Ochsner

5.2 Ground conditions

In positioning the horizontal collector or a vertical loop, it is important that the ground is well settled and level. Otherwise, the piping system could be damaged if the ground shifts. The more water the ground contains, the better the heat transmission. A smaller ground area is required for dense, wet soil than for dry crumbling ground.

Any area disturbed when installing the necessary collector area may be re-landscaped. Rainwater is very important for the regeneration of the ground. Standing water or flooding should be avoided (slopes or clay-earth), as ground swelling and damage may occur. If necessary, proper drainage should be installed.

The ground must settle before heat is collected. Limited ice build-up around the collector tubing is allowable and even preferred (for better heat transfer).

5.3 Layout and installation of ground collector

The size of the heat collector depends on the capacity of the heat pump (model, expected COP) and the specific heat collection capacity of the ground. The collector size also needs to take account of the total annual heating demand which, for domestic heating operation, is typically between 1700 and 2300 hours in central Europe (see Tables 5.1, 5.2 and 5.3).

COLLECTION CAPACITY (COOLING CAPACITY WHEN HEATING) = HEATING CAPACITY – POWER CONSUMPTION (ELECTRIC)
This method of sizing is also appropriate for seasonal cooling service.

5.3.1 Horizontal collector installations

INSTALLATION AREA
The size of the area for the collector can be calculated with the help of Table 5.1.

Normally, the installation location is evenly flat, or with a slight, consistent slope. In sloping systems, the collector pipes must be laid perpendicular to the slope.

Care should be taken to properly air-bleed the piping system.

Caution: An undersized heat pump (electrical element required) and/or undersized collector or loop can lead to permanent icing, distortion of the ground and increased heating costs.

INSTALLATION SPACING
This is calculated using the total collector area required and the total collector tube length. An approximate value for a normal collector in damp, well packed soil is at least 50cm. In dry, sandy, loose soil, at least 80cm.

Table 5.1 Heat collection capacity

Ground Conditions	Specific Heat Collection Capacity	Collector Area per kWth		
		at β = 3	at β = 3.5	at β = 4
Dry, loose soil	10 W/m²	66 m²	71 m²	75 m²
Damp, packed soil	20 – 30 W/m²	33 – 22 m²	36 – 24 m²	38 – 25 m²
Saturated Sand/Gravel	40 W/m²	17 m²	18 m²	19 m²

β = SPF = Seasonal Performance Factor; with approx. ß = 4 assumed   Source: VDI 4640

INSTALLATION DEPTH
Typical installation depth in Europe varies between 0.8 and 1.5m. Consideration should be given to local frost depth and extent of snow cover in winter.

INSERTION
It is important to ensure that the pipe is not crimped or squeezed. The collector pipes should be laid on and covered with a protective layer of sand before covering with earth.

Exception: The soil has characteristics similar to that of sand. In this case, the burial process must be carried out with great care. The pipe should be completely covered (ground level) and soil replaced in the proper location (without damaging collector pipes).

COLLECTOR PIPE
PE-Pipe Hard PN 10 (25•2,3=0.75"=DN 20 or 32•3=1"=DN 25). It is important that the pipe sections are of the same length in order to achieve nearly equivalent

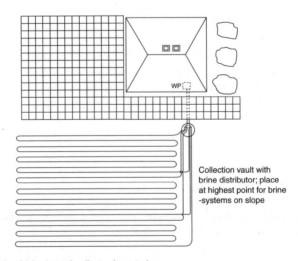

Collection vault with brine distributor; place at highest point for brine -systems on slope

Figure 5.2 Example of a horizontal collector layout plan
Source: Ochsner

pressure differences (for even circulation and energy collection). An exact schematic of the collector pipe network and the installation area should be produced. Photos and drawings are recommended. The final layout of the collector should minimize pressure drop while ensuring turbulent flow.

WARNING STRIP
It is recommended to place warning strips, or tapes, approx. 50cm above the collector pipes.

5.3.2 Trench collector/spiral collector
Spiral trench collectors are recommended when a small area is available.

Table 5.2 Heat collection capacity/spiral collector

Ground Conditions	Specific Heat Collection Capacity
Damp, packed soil	100 to 125 W/m trench

Installation depth	1.6 to 2.0m
Trench width	min. 80cm
Trench length	min. 20 m per loop (up to 30m) with 125m tube
Collector pipe	PE-Pipe Hard PN 10 (32 ×3), 125m per loop (up to 200m)
Insertion	as in 5.3.1 Horizontal Collector (Adhere to workplace safety practices)

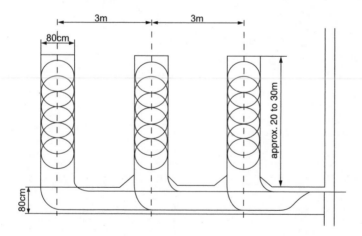

Figure 5.3 Trench collector with connection trench
Source: Ochsner

5.3.3 Vertical loop
Vertical loops require the smallest installation area.
   The temperature at a depth of about 15m remains nearly constant all year round.

Below 100m, the temperature of the earth increases with increasing depth (approx. 1°C per 30m).

The following table gives the values for double U configuration.

Table 5.3 Heat collection capacity

Ground Conditions	Specific Heat Collection Capacity	Loop length per kW th		
		at ß = 3	at ß = 3.5	at ß = 4
Dry, sediment	30 W/m	22.0 m	24.0 m	25.0 m
Shale, Slate	55 W/m	12.0 m	12.0 m	14.0 m
Solid stone with high thermal conductivity	80 W/m	8.0 m	9.0 m	9.5 m
Underground with high groundwater flow	100 W/m	6.5 m	7.0 m	7.5 m

ß = SPF = Seasonal performance factor; with approx. ß = 4 assumed   Source: VDI 465047

Figure 5.4 depicts the temperature profile 10m from the vertical loops, each on the first of the month at noon.

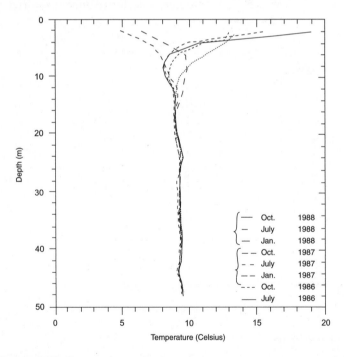

Figure 5.4 Temperature Profile
Source: Professor B. Sanner, University of Giessen

Loop depth: depending on system, usually around 100m
Loop spacing: Should be separated by at least 5m

DRILLING

Geological analysis should be completed before drilling. This will give an indication of underground layers and an exact collection capacity. The drilling is to be

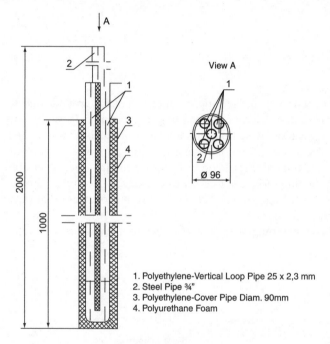

1. Polyethylene-Vertical Loop Pipe 25 x 2,3 mm
2. Steel Pipe ¾"
3. Polyethylene-Cover Pipe Diam. 90mm
4. Polyurethane Foam

Figure 5.5 Construction of a typical double-u earth loop (0.75" up to 50m)
Source: Ochsner

Figure 5.6 Drilling for vertical loop collectors is to be completed by competent drilling companies
Source: Ochsner

completed by a specialized and licensed drilling company. If the geological conditions are unclear, a test drilling is useful. The insertion of the loop pipe is also completed by the drilling company. The hole will be refilled and the tubing secured according to industry standard, to include proper sealing in the case of ground water. Temporary casing may be required in loose geological formations.

The static pressure must be taken into account when carrying out the pressure test.

5.4 Connection

5.4.1 Collection vault

It is recommended to bring the supply and return of the ground collector together in a collection vault outside of the building (see Figure 5.7). With many collector systems, especially those which are separated by more than 10m from the heat pump, a collection vault is of absolute necessity. They can easily be constructed using commonly found cement rings. In order to ensure adequate access, clearances of less than 150mm should not be used in any case. The collector tubes should be sloped up toward the distributor and accumulator in order to ensure that the system can be easily bled.

The connection piping from the building to the collection vault should be

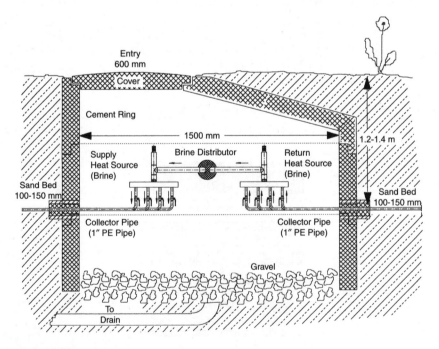

Figure 5.7 Collection vault
Source: Ochsner

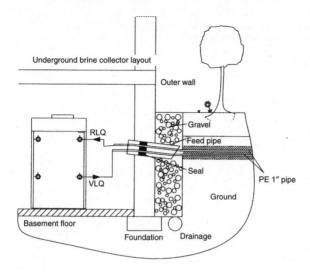

Figure 5.8 Wall feed-through
Source: Ochsner

straight and sloping down slightly in the direction of the collection vault. This allows for falling condensate to be drained away. The connection pipes should be properly insulated, and installed in plastic ductwork if necessary.

5.4.2 Safety clearance
The piping must be installed at least 1.5m away from water pipes and 1m away from electrical cables. If layout is parallel to a building, it should be separated by approximately 1.2m (to avoid frost damage).

If these safety clearances are not met, special steps must be taken: insulation of the object or the collector pipe with closed cell insulation.

If the pipe is to be encroached upon, the pipe should be properly protected and well insulated (closed cell insulation). The pipe should also be well insulated at the building entry point.

5.4.3 Building penetrations
The wall penetration as shown in Figure 5.8 is constructed using ductwork which slopes down slightly toward the outside. It should be sealed with foam or if necessary a special sealing compound. A drainage system should be installed to ensure that no water enters the building in the event of heavy rainfall.

5.5 Brine circulation loop

The brine circulation loop (Figures 5.9 and 5.10) normally consists of the following components: collector pipes, manifold, vent, circulation pump, expansion vessel, safety valve, insulation or condensate drainage, and flexible connections to the heat

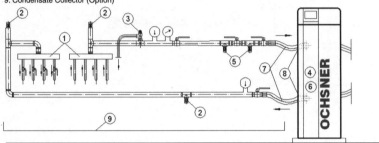

Hydraulic Schematic – Heat Collector System (without collection vault)
For Brine/Water Heat Pump GMSW (Golf-Midi/Maxi)

1. Brine Distributor
2. Manual Air-bleed Valve
3. Safety Valve
4. Expansion Tank
5. Ball Valve
6. Brine Circulation Pump
7. Flexible Connection Hose
8. Supply/Return Heat Source
9. Condensate Collector (Option)

Integrated in GMSW 6-21 Heat Pump

Pipes, flexible connections, and fixtures should be insulated with vapor diffusion tight insulation.
The outer thread of screw joints should be wrapped in Teflon tape to prevent the entry of condensates.

Figure 5.9 Brine system with distributor in installation room (integrated circulation pump in heat pump)
Source: Ochsner

pump (closed systems). Manifolds which allow for the individual isolation of each brine loop are recommended. If possible, continuous pipe loops to and from the manifold avoids the need for welding in underground areas.

All components – including the flexible hose – should be corrosion free materials and should be insulated in the heating room to prevent condensation – using closed cell insulation.

The piping should be sized such that the fluid velocity does not exceed 0.8m/s.

SIZING OF THE BRINE CIRCULATION PUMP
The brine mass flow rate must be capable of transporting the full thermal capacity required from the heat source.

Required for the determination of the flow rate are:

■ Cooling capacity (= heat collection capacity)

■ Temperature difference, heat collection supply/return ($\Delta t = 3K$)

■ Specific heat capacity of brine.

Values of the minimum flow rate for each heat pump model are usually provided in the manufacturer's data sheets.

$$\dot{m}_b = \frac{P_K \cdot 3600}{c \cdot \Delta t} \text{ (kg / h) Mass flow rate}$$

(Corresponds approx. to flow rate in l/h.)

P_K = Cooling capacity = heat collection capacity (kW)
c = Specific heat capacity of brine = 3.9 kJ/kg K
(for 30 per cent propylene glycol solution)*
(conversion factor 1kWh = 3,600 kJ)
Δt = temperature difference (e.g. 3 K)

Required for the determination of the required pressure head are:
– Head loss in collector pipes (from pipe roughness charts)
– Head loss in heat pump evaporator (from "Heat Pump Technical Data")
– Head loss of fixtures (approx. 50 mbar)
– Correction for increased viscosity due to anti-freeze (guideline: 1.3 to 1.5 bar)

The ground loop circulation pump must be sized to achieve the following:
– minimum flow rate through the heat pump

* 30 per cent propylene glycol needed in cold climates only. Other options like ethylene glycol are possible.

Hydraulic Schematic – Heat Collector System (with collection vault)
For Brine/Water Heat Pump GMSW (Golf-Midi/Maxi)

1. Feed Pipe
2. Manual Air-bleed Valve
3. Safety Valve
4. Expansion Vessel
5. Drain
6. Brine Circulation Pump
7. Flexible Connection Hose
8. Supply/Return Connections
9. Condensate Collector (Option)

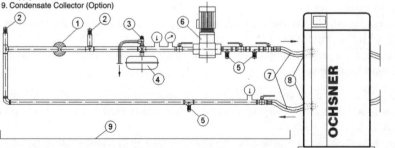

Integrated in GMSW 6-21 Heat Pump

Pipes, flexible connections, and fixtures should be insulated with vapor diffusion tight insulation.
The outer winding of screw joints should be wrapped in Teflon tape to prevent the entry of condensates.

Figure 5.10 Brine system with distributor in collection vault (circulation pump not integrated in heat pump)
Source: Ochsner

– turbulent flow (Reynolds number > 2100) under the worst (coldest) flow conditions
– minimium head loss and hence pumping power.

Figure 5.10 shows a brine system with the distributor in a collection vault and the circulation pump not integrated in the heat pump.

5.6 Commissioning

INSTALLATION

Installation of a heating heat pump which uses a safety refrigerant is possible in any room which is both dry and protected from freezing temperatures. The system should be installed on an even, horizontal surface. The construction of a free-standing base is recommended. Placement of the unit should be such that servicing and maintenance are possible. Generally, only flexible connections to the heat pump should be implemented.

For heat pumps which use inflammable refrigerants, technical and safety guidelines and codes are to be followed (ventilation, explosion protection, fire protection doors etc. from standards and manufacturer data).

LEAK CHECK

All system components, collector and loop must be tested for the presence of leaks. The results of the tests should be documented. Ground loop arrays should be pressure tested prior to being backfilled to allow for remedial works in the event of leaks being detected.

FILLING AND AIR-BLEEDING

When filling the brine system, the following procedure should be followed:

1. Preparation of the anti-freeze mixture using a large storage container is recommended before filling.
2. Flush ground loops clean to remove contaminants. (For 100m vertical loop, at least 3 min. at 2 to 3 bar; use a fill pump or use tap water pressure if it is adequate.)
3. Fill with fill pump with sufficient capacity: avoid aerating the water.
4. Circulate for long enough to ensure that all air has been bled from the complete ground loop array.

The selection of freeze protection and brine composition is to occur according to the manufacturer's data. The system is to be thoroughly bled of air to ensure proper operation.

Check velocity in the ground loop collectors in order to ensure proper heat-transfer.

COMMISSIONING

Commissioning should be performed by factory or distributor customer service or qualified representatives. The heat collection, heat distribution and electrical connections must all be in operating order before the time of commissioning.

A commissioning protocol will be followed and a system operation orientation will be given.

5.7 CO$_2$ loop

See Chapter 14.8.

6

Planning Instructions for Ground Heat Source – Direct Expansion Systems

6.1 Ground heat source

The ground serves as an ideal heat source for monovalent operation systems. The ground stores solar energy and is also regenerated with rainwater. As a result, sufficient energy is available in winter, even when the ground is covered in snow.

SYSTEM DESCRIPTION (see Figure 6.1)
In direct expansion system, the refrigerant in the heat pump is circulated through the ground collector, where it undergoes vaporization. This direct expansion enables the highest performance coefficients to be achieved and provides the greatest operation reliability, as heat exchanger and brine circulation pump can be eliminated. A horizontal collector heat collection system is used. The collector is also referred to as the vaporization loop. The typical loop is 75m long and is constructed of seamless copper tubing with a polyethylene sheath. The method of installation varies but must adhere to any applicable standards. In some cases, permits must be obtained. This recognized technology is also described in the DIN 8901.

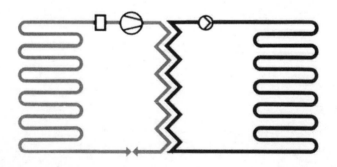

Figure 6.1 Simplified Operating Scheme Direct Expansion/Water
Source: Ochsner

6.2 Ground conditions

In positioning the direct expansion collector, it is important that the ground is well settled and level, as the piping system could be damaged if the ground shifts. The more water the ground contains, the better the heat transmission. A smaller ground area is required for dense, wet soil than for dry crumbling ground.

Any area disturbed when installing the necessary collector area may be re-landscaped. Rainwater is very important for the regeneration of the ground. Standing water or flooding should be avoided (slopes or clay soil), as ground swelling and damage may occur. If necessary, proper drainage should be installed.

The ground must settle before heat is collected.

6.3 Layout and installation of collector

The size of the heat collector depends on the capacity of the heat pump (model, expected COP) and the specific heat collection capacity of the ground. The collector size also needs to be sufficient to meet the total annual heating demand. The heat requirement for domestic heating operation is typically 1700 to 2300 hours annually in central Europe. The ground collector needs to take account of this annual heating demand (see Table 5.1, 5.2 and 5.3).

Collection capacity (cooling capacity when heating) = heating capacity – power consumption (electric)

This method of sizing is also appropriate for a seasonal cooling service.

6.3.1 Horizontal loop installations

INSTALLATION AREA
The size of the area for the collector is calculated with the help of Table 6.1.

Normally, the installation location is level and even, or with a slight, consistent slope. In sloping systems, the manufacturer's instructions are to be followed.

Table 6.1 Heat collection capacity

Ground Conditions	Specific Heat Collection Capacity	Collector Area per kWth		
		at $\beta = 3$	at $\beta = 3.5$	at $\beta = 4$
Dry, loose soil	10 W/m²	66 m²	71 m²	75 m²
Damp, packed soil	20 – 30 W/m²	33 – 22 m²	36 – 24 m²	38 – 25 m²
Saturated Sand/Gravel	40 W/m²	17 m²	18 m²	19 m²

ßa = SPF = Seasonal Performance Factor; with approx. ßa = 4 assumed   Source: VDI 4640

INSTALLATION SPACING

This is calculated using the total collector area required and the total collector tube length. An approximate value for a normal collector in damp, well-packed soil is at least 50cm. In dry, sandy, loose soil, at least 80cm.

INSTALLATION DEPTH

Typical installation depth in Europe varies between 0.8 and 1.5m. Consideration should be given to local frost depth and extent of snow cover in winter.

INSERTION

It is important to ensure that the pipe is not crimped or squeezed. The collector pipes should be laid on and covered with a protective layer of sand before covering with earth. The burying process must be carried out with great care. The pipe should be completely covered (ground level) and soil replaced in the proper location (without damaging collector pipes).

COLLECTOR MATERIAL

In the field, copper collector tubing has performed well. It is used in the USA, Canada and Scandinavia, unsheathed. In Austria and Germany, standards are more stringent and only seamless tubing with a plastic sheath is used. The sizing is accomplished according to manufacturer' data.

COLLECTOR LENGTH

Normally, the full 75m length of the vaporization loop must be used for heat collection. The loop may only be shortened with the approval of the manufacturer's customer service department.

WARNING STRIP

It is recommended that warning tapes are placed approximately 50cm above the collector pipes.

REFRIGERANT/REFRIGERANT OIL

With the use of a chorine-free safety refrigerant, biodegradable, synthetic refrigerant oil (Esther oil) is used, thus eliminating another possible environmental threat. According to the VDI 4640 Guideline, R 407C is classified as being the same as normal brine antifreeze with WGK=1.

FURTHER SAFETY MEASURES

Direct expansion systems contain an extra solenoid valve for additional safety. In the case of pressure loss in the collector, the low pressure safety switch signals for the collector loop to be closed. By closing the solenoid valve, no additional refrigerant from the heat pump can enter the collector loop.

6.4 Connection

6.4.1 Collection vault

With nearly all collector systems, especially those in which the collector does not connect directly to the wall feed-through, a collection vault is necessary. They can easily be constructed using commonly found cement rings. In order to ensure adequate access, clearances of less than 150mm should not be used in any circumstances.

An insulated feed pipe (a pipe chase or duct) should be used for the connection between the building and the collection vault. The feed pipe should be straight and angled down slightly in the direction of the collection vault, allowing for the drainage of any condensate.

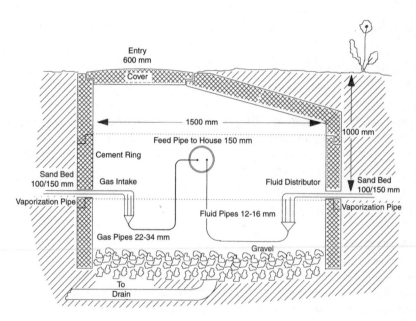

Figure 6.2 Collection vault – direct expansion
Source: Ochsner

6.4.2 Safety clearance

The piping must be installed at least 1.5m away from water pipes and 1m away from electrical cables. If the layout is parallel to a building, it should be separated by approximately 1.2m. (Avoid frost damage!)

If these safety clearances are not met, special steps must be taken: insulation of the object or the collector pipe with closed cell insulation. If the pipe is to be encroached upon, the pipe should be properly protected and well insulated (closed cell insulation). The pipe should also be well insulated at the building entry point!   <AQ2>

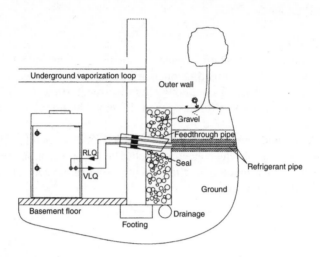

Figure 6.3 Wall feed-through
Source: Ochsner

6.4.3 Building penetration

The wall penetration as shown in Figure 6.3 is to be constructed using a pipe duct which slopes down slightly towards the outside. It should be sealed with foam or if necessary with a special sealing compound. A drainage system should be installed to ensure that no water enters the building in the event of heavy rainfall.

6.5 Commissioning

INSTALLATION

Installation of a heating heat pump which uses a safety refrigerant is possible in any room which is both dry and protected from freezing temperatures. The system should be installed on an even, horizontal surface. The construction of a free-standing base is recommended. Placement of the unit should be such that servicing and maintenance are possible. Generally, only flexible connections to the heat pump should be implemented. All piping carrying refrigerant should be insulated in order to prevent condensation.

In cases of outdoor installation, proper measures which allow for service and maintenance, as well as providing corrosion and frost protection, are to be taken. In any event, it is difficult to avoid frost damage in the event of power outages.

COMMISSIONING

The vaporization loop is to be connected and filled with refrigerant only by the manufacturer's customer service department. The heat distribution and electrical connections must all be prepared for operation before the time of commissioning. A commissioning protocol will be handed over to the user and a system operation orientation should be made.

7

Planning Instructions for Water Heat Source Systems

7.1 Water heat source

Water/water heat pumps utilize energy from groundwater, certain surface water, or from water in heat rejection systems.

With groundwater as the heat source, heat pumps can achieve the highest performance coefficients. Groundwater remains at a nearly constant temperature of 8 to 12°C throughout the year in moderate climates. Water temperatures can be higher or lower depending on the local climate zone. The result is that, in comparison to other heat sources, the temperature increase required for useful heating purposes is relatively small. The groundwater table should not be deeper than 15m to limit pipework friction losses.

SYSTEM DESCRIPTION (see Figure 7.1)
The heat source system consists of a supply well with a submerged pump and a reinjection well into which the water flows after being circulated through the heat pump. The groundwater is cooled about 4 K in the heat pump. The groundwater heat

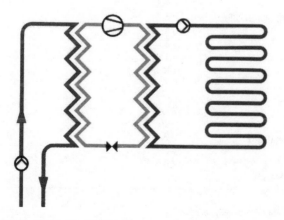

Figure 7.1 Simplified operating scheme water/water
Source: Ochsner

source system is generally an open system. Any necessary permits must be obtained from the responsible agency before beginning any construction.

7.2 Ground and well conditions

Near surface groundwater may be used. The adequacy of the volume and temperature of the groundwater is to be tested by three days of continuous pumping operation – or as required by the local water/licensing authority. The groundwater temperature should not fall below 8°C, in order to prevent damage to the evaporator.

In order to avoid damage due to corrosion, the conductivity of the water should not exceed 450 micro Siemens per cm. Water analysis is recommended. Caution: water quality may change with time – for example, due to local fertilizer use.

The use of a submerged pump avoids the possibility of introducing air or oxygen into the system. A back-washable filter should also be installed.

Table 7.1 shows non-binding recommendations. Please use manufacturer data for concrete design values.

Table 7.1 Water quality

Electric Conductivity (μ-Siemens/cm*)	> 600	−
pH-value	< 6	0
	6 to 8	+
	> 8	−
Chloride (mg/l)	< 10	+
	10 to 100	+
	100 to 1000	0
	> 1000	−
Sulfate (mg/l)	< 50	+
	50 to 200	0
	> 200	−
Nitrate (mg/l)	< 100	+
Carbonic Acid (mg/l) (abrasive/free)	< 5	+
	5 to 20	0
	> 20	−
Oxygen (mg/l)	< 1	+
	1 to 8	0
	> 8	−
Ammonium (mg/l)	< 2	+
	2 to 20	0
	> 20	−
Iron and Manganese (mg/l)	> 0.2	0
Sulfide (mg/l)	< 5	−
Chlorine (free) (mg/l)	< 0.5	+
Allowable Organic Material		0

* At 20°C
+ = Construction material is usually resistant
0 = Corrosion accumulation of mud and sedimentation of iron ochre can occur when multiple factors are given a '0' rating
− = Implementation is discouraged
Source: Ochsner

OTHER CONFIGURATIONS

Open water, industrial exhaust heat etc. can be used as indirect heat sources through the implementation of an intermediate loop.

7.3 Design

The wells can be dug or drilled (minimum diameter 220mm required for the downhole pump) and should be constructed by an experienced company. In finely granulated earth, the diameter must be increased in order to prevent entry of sand into the flow. Impact drilling should be avoided.

The return well should be located at least 10 to 15m in the downstream direction of the groundwater flow.

SIZING THE WELL PUMP

The flow rate of water must be capable of delivering the full capacity required from the heat source.

(Approximate value: 160 litre/h per kW heat pump heating capacity)

Required for the determination of the flow rate are:

■ Cooling capacity (= heat collection capacity)

■ Temperature difference, heat collection supply/return ($\Delta t = 4K$)

■ Specific heat capacity of water.

Values for flow rate for each model can also be seen in 'Heat Pump Technical Data of manufacturer.

$$\dot{m}_w = \frac{P_K \cdot 3600}{c \cdot \Delta t} \text{ (kg / h) Mass flow rate}$$

(Corresponds approximately to flow rate in l/h)

P_K = Cooling capacity = heat collection capacity (kW)
c = Specific heat capacity of water = 4.2 kJ/kg K
(Conversion factor 1 kWh = 3600 kJ)
Δt = temperature difference (e.g. 4 K)

Required for the determination of the required pressure head are:

■ Geostatic pressure head (the water level difference is relevant during the operating conditions, as well as for planning the required pressure head)

■ Head loss in collector pipes (from pipe roughness charts)

■ Head loss in heat pump evaporator (from "Heat Pump Technical Data")

■ Head loss of fixtures (approx. 2m liquid column)

7.4   Connection

The connection piping should be angled downwards towards the wells and be adequately insulated against freezing. The flow velocity should not exceed 0.8m/s. All groundwater pipes, including flexible hoses in the house, should be insulated to prevent condensation.

An additional feed pipe between the heat pump and the supply well is required for the electric cable to the downhole pump (see Figure 7.2).

7.5 Components/filter

The heat source system consists of the following components (see diagram): supply and return wells, well pump (usually submerged), check valve, piping network, filter (back-washable), two manometers, two thermometers, and flexible connections to the heat pump.

7.6 Building penetrations

The wall penetration is to be completed in the manner previously described in Chapter 6.4.3.

7.7 Commissioning

INSTALLATION
Installation of a heating heat pump which uses a safety refrigerant is possible in any room which is both dry and protected from freezing temperatures. The system should be installed on an even, flat surface. The construction of a free-standing base is recommended. Placement of the unit should be such that servicing and maintenance are possible. Generally, only flexible connections to the heat pump should be implemented.

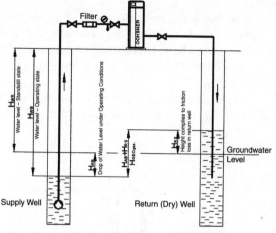

$H_{ges} = H_{tot}$ pressure head of system = $H_{AB} + H_{RS} + H_{WP} + H_{RR} + H_{AR}$

H_{AB} = Drop of water level under operating conditions ($H_{WB} - H_{WR}$)
H_{RS} = Friction Losses in Return Well (rule of thumb: 400 mbar or 4 liquid column)
H_{WP} = Pressure Difference in Heat Pump Evaporator
H_{RR} = Pipe Friction Losses
H_{AR} = Component Pressure Losses (Valves and Filter)

Connection Diagram for Heat Collection System (Well)
with Water/Water Heat Pump GMWW (Golf Midi/Maxi)

1. Submerged Pump (with non-return-valve)
2. Water Filter
3. Drain
4. Feed Pipe
5. Heat Source Supply, Return
6. Flexible Connection Hose
7. Water Flow Gauge

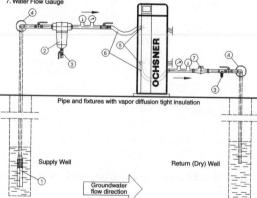

Figure 7.2 Well system
Source: Ochsner

COMMISSIONING

Commissioning is to be performed by the manufacturer's customer service department or qualified representatives. The heat collection, heat distribution and electrical connections must all be in working order before the time of commissioning. A commissioning protocol will be handed over to the user and a system operation orientation should be made.

The drill diameter should be at least 220mm (for sandy conditions, more, in order to prevent sand entry), Figure 7.3 shows the construction of a drilled well.

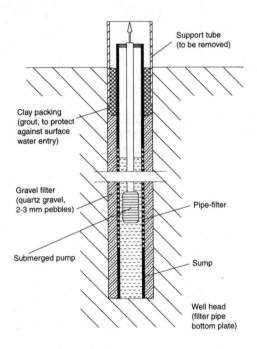

Figure 7.3 Drilled-Well Construction
Source: Ochsner

Warning: In incorrectly constructed or undersized wells, the system may become clogged with sand!

8

Planning Details for Air Heat Source Systems

8.1 Air heat source

Outdoor air is a heat source which is available anywhere and can be used without special permits. When outdoor temperatures fall, the heating requirements for a building increase. At the same time, the heating capacity and coefficient of performance of the heat pump decrease. With an integrated automatic defrost function, air/water heat pumps operate in temperatures as cold as –15°C or below.

SYSTEM DESCRIPTION (see Figure 8.1)
Outdoor air is blown with a fan over the heat pump evaporator, which draws energy out of the air as it passes.

With compact systems the evaporator is combined with the other elements of the heat pump into a single enclosure. The built-in fan must overcome the pressure losses of the supply and return ducts required for indoor installation.

In split systems, the evaporator is installed outdoors. Due to the unrestricted air flow, extremely quiet axial fans can be used. The connection consists of refrigerant lines; air ducts are eliminated. With the use of a safety refrigerant, systems can be installed in any location.

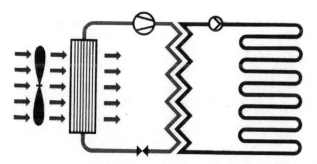

Figure 8.1 Simplified operation scheme air/water
Source: Ochsner

8.2 Design

8.2.1 Split system

Split systems combine the advantages of outdoor and indoor installation. No ducts are required and the space-saving heat pump is protected within the building. The split evaporator should be equipped with a quiet operating, slow rotating axial fan. Only two refrigeration lines need to be connected through the building wall and insulated.

INDOOR UNIT INSTALLATION

Installation of an indoor unit with a safety refrigerant is possible in any room which is both dry and protected from freezing temperatures. The system should be installed on an even, horizontal surface. The construction of a free-standing base is recommended. Placement of the unit should be such that servicing and maintenance are possible. We recommend the following clearances: distance from rear-side of heat pump to wall minimum 50cm, distance from sides to wall minimum 40cm, distance from front-side to wall minimum 70cm.

OUTDOOR UNIT INSTALLATION

The distance from the heat pump to the split evaporator unit can be many metres (see manufacturers' data). The evaporator should be equipped with a defrost mechanism. Frost protected drainage for condensate must be provided. In order to ensure good drainage, a drain trough should be installed on site. If no such collector is possible, the unit should be placed on a gravel bed. In winter, ice build-up may occur around the drain outlet.

Care should be taken to prevent the outlet air from freezing paths or walkways.

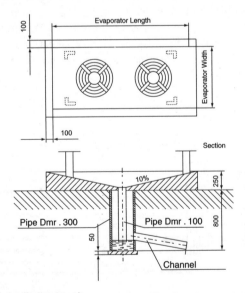

Figure 8.2 Recommended collection trough
Source: Ochsner

NOISE EMISSION

The sound power emitted by the unit is constant.

Emission levels are a measure of sound pressure level at a set distance from the device. These distances are set out in various local standards.

According to VDI 2085, the following values for a neighbour with open windows may not be exceeded (Sound pressure level day/night):

– Commercial/residential areas 60 dB(A)/45 dB(A)
– Normal residential areas 55 dB(A)/40 dB(A)
– Exclusively Residential Areas 50 dB(A)/35 dB(A)

The sound pressure level measured at 1m from heat pumps in an open area is approximately 8 dB under the sound power level. The sound pressure level falls an additional 2 dB(A) with each metre of distance.

The following points should be noted when positioning the evaporator (outdoor) unit:

– Avoid installation on reverberating surfaces (plaster, cement, asphalt etc.)
– Installation between two walls or in a corner may increase the sound emissions
– Avoid installation of the unit near bedrooms
– Plants and vegetation can reduce sound emissions.

PIPING CONNECTION TO EVAPORATOR

As shown in Figures 8.3 and 8.4, the connection lines to the outdoor unit (Liquid and Vapour), as well as the electrical lines to the fan, should be installed in a conduit. The connecting lines must be those provided by the manufacturer, or by a qualified refrigeration technician.

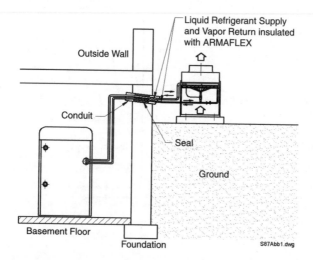

Figure 8.3 Piping connection directly to evaporator
Source: Ochsner

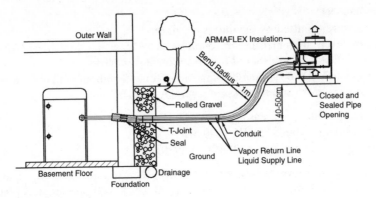

Figure 8.4 Underground piping connection
Source: Ochsner

8.2.2 Compact systems – indoor set-up

INSTALLATION

Installation of an indoor unit with a safety refrigerant is possible in any room which is both dry and protected from freezing temperatures. Indoor set-up simplifies maintenance, protects the unit from weathering and provides freeze protection in the event of power outage.

The system should be installed on an even horizontal surface. In addition to the built-in condensate drain, a drainage channel (Figure 8.5) should be constructed under the unit to collect any possible condensation (avoidance of water damage).

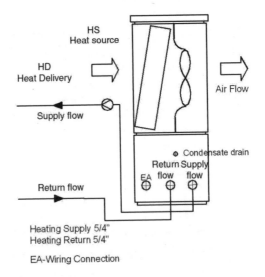

Figure 8.5 Set-up of compact unit
Source: Ochsner

Placement of the unit should be such that servicing and maintenance are possible. Appropriate clearance is recommended.

If the heat pump is set up in a closed room where combustion occurs using air from the room, an additional ventilation opening is necessary. Leaks in the air intake opening would lead to unacceptable drops in the air pressure in the room.

AIR DIRECTION

The direction of air is achieved with proper air ducts and is dependent upon the air flow rate of the heat pump (see Figures 8.6–8.8).

The intake air should not contain corrosive contaminants (ammonia, sulphur, chlorine etc.).

8.2.3 Compact systems – outdoor set-up

Compact outdoor units can be set up in any open location. The air intake and exhaust should be unrestricted. All components of the refrigeration cycle are integrated into the unit, so no air ducts are required. The housing must provide protection against weather elements. The connection to the heating system in the house consists of two insulated pipes for the supply and return. These are to be installed underground with the heat pump power supply and control cable. Noise emissions are dependent upon sizing, construction and speed of the fan.

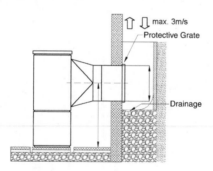

Figure 8.6 Air intake or exhaust via light well
Source: Ochsner

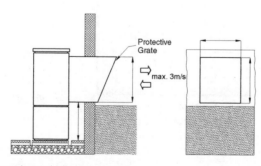

Figure 8.7 Air intake or exhaust via direct wall opening
Source: Ochsner

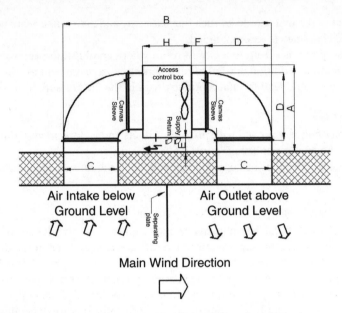

Figure 8.8 Air connections and ducts are to be insulated against condensation
Source: Ochsner

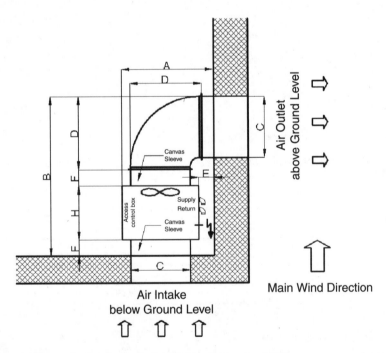

Figure 8.9 Recommended placement of air intake and exhaust on building
Source: Ochsner

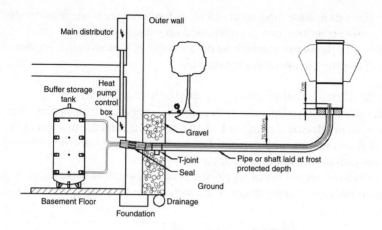

Figure 8.10 Connection for outdoor set-up
Source: Ochsner

To avoid noise disturbances caused by poor unit placement, follow manufacturer's instructions regarding orientation.

The defrost function allows for outdoor temperatures down to approximately −16°C (various systems).

Sound emissions – see previous comment.

8.3 Operation configuration

8.3.1 Monovalent systems (max. supply temp. 55°C to 65°C)

For monovalent operation, a heat pump with an integrated defrost mechanism is required. The heat pump must be sized such that the heating capacity meets the heating demand at a specified operating condition (the design temperature). Because the heating capacity of an air/water heat pump at an outdoor temperature of 15°C

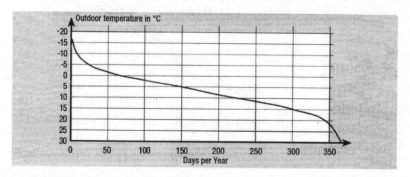

Figure 8.11 Annual load duration curve (Austria, Bavaria)
Source: AWP

is three times its heating capacity at –15°C, such systems are made considerably over-sized to cover transition periods (increased investment costs).

Figure 8.11 shows the annual load duration curve. This gives the number of days on which a certain temperature is exceeded.

8.3.2 Bivalent-parallel (max. supply temp. 55°C to 65°C)

The heat pump heats independently to a certain set point, at which point an auxiliary heating system (electric element or boiler) is turned on and the two systems operate in parallel to meet heating demand (monoenergetic with electric element). (For determination of set point, see Chapter 4.4.5.)

The low investment cost of an electrical heating element makes it an attractive system variation. If a supply temperature higher than 65°C is required, the electrical

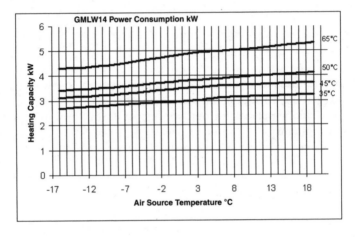

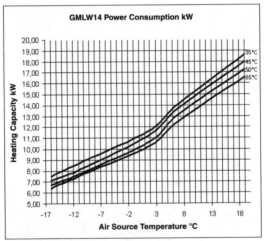

Figure 8.12 Example of a performance curve (Ochsner GMLW 14 plus)
Source: Ochsner

element must be sized to match the full system demand. However, care must be taken to ensure that excessive use of direct electric heating elements does not give rise to unacceptably high running costs or carbon emissions.

Heat pumps operating in this configuration must also be equipped with a defrost mechanism.

8.3.3 Bivalent-alternate (suitable for supply temperatures up to 90°C)

The heat pump heats independently to a certain set point. Once this point is reached, a boiler meets the full heating demand. (For determination of set point, see Chapter 4.4.4.)

8.4 Commissioning

The defrost mechanism must be properly set during commissioning. This is to be completed by the manufacturer's customer service department or a qualified refrigeration technician.

In Figure 15.5, a diagram of a mono-energetic system is shown. The electric heating element is placed in the buffer tank.

9

Planning Instructions – Heating Systems

The heating system (Heat Distribution) consists of the following components:

■ buffer/hydraulic separator storage tank or low loss header

■ circulation pump

■ piping network and connection set

■ heat distribution system: radiant floor/wall surfaces, radiators

■ possibly domestic hot water heating

For recommended hydraulic schemes, see Chapter 15.

9.1 Heating circulation loop control

The control of the heating loop should enable a variable supply temperature from the heat pump. A mixing valve leads to increased supply temperatures from the heat pump and a corresponding decrease in the coefficient of performance. Generally, only systems which make use of both radiators and radiant floor heating require the use of a mixing valve.

9.2 Hydraulic separation

It should be ensured that the full heating capacity of the heat pump will be consistently used. This is achieved with a constant, sufficient flow rate in the heating system. In order to ensure this during all operating conditions, the installation of hydraulic separation is recommended. This can be done using a buffer tank (Figure 9.1) or low loss header (Figure 9.2).

This also serves to prevent frequent on/off cycling of the compressor (reducing lifespan and providing the heating system with only intermittent heat).

**Hydraulic Schematic of Radiant Floor Heating System
(with buffer tank) with GMSW heat pump**

1. Buffer circulation (feed) pump
2. Safety valve
3. Drain
4. Expansion vessel
5. Manual air-bleed-valve (not automatic)
6. Heating supply/return
7. Radiant floor heating distributor
8. Parallel buffer tank (30 l/kW)
9. Heating circulation pump
10. Flexible connection hoses
11. Check valve

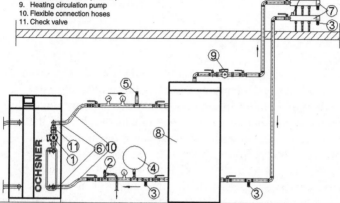

Pipes, flexible connections, and fixtures should be insulated with vapor diffusion tight insulation.
The outer winding of screw joints should be wrapped in Teflon tape to prevent the entry of condensates.

Figure 9.1 Heating system with buffer tank for hydraulic separation and buffer effect
Source: Ochsner

**Hydraulic Schematic of Radiant Floor Heating System
(with low loss header) with GMSW heat pump**
1. Buffer circulation (feed) pump
2. Safety valve
3. Drain
4. Expansion vessel
5. Manual air-bleed-valve (not automatic)
6. Heating supply/return
7. Radiant floor heating distributor
8. Low loss header
9. Heating circulation pump
10. Flexible connection hoses
11. Check valve.

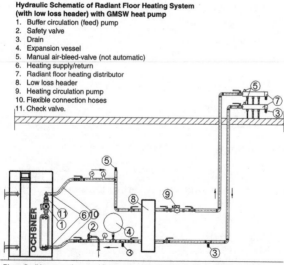

Pipes, flexible connections, and fixtures should be insulated with vapor diffusion tight
insulation.
The outer winding of screw joints should be wrapped in Teflon tape to prevent the entry
of condensates

Figure 9.2 Heating system with low loss header for hydraulic separation
Source: Ochsner

For increased heating comfort, for control of multiple heating zones (to account for solar gains etc.), or for the independent shut-off of certain areas, the inclusion of hydraulic separator, at least a low loss header, is necessary. Alternatively, the use of a bypass valve is possible.

With proper sizing, the hydraulic separator can also serve as a buffer tank (see below).

9.3 Buffer storage tank

If a buffer/storage effect is necessary, a normal buffer storage tank should be installed (to balance loads, extend operating cycle length and/or allow for heating with radiators during utility interruption periods). With unzoned radiant floor heating, the use of a buffer storage tank is not required, as the thermal mass of the floor installation stores enough warmth to bridge over interruption times.

The simplest hydraulic arrangement is to use the buffer storage as a low loss header by installing it in a parallel configuration. It is often possible to connect the buffer storage tank in series with the heating supply (with implementation of an electric heating element) or in the heating return (follow manufacturer's guidance).

BUFFER SIZING
As a guideline for buffer storage sizing, a value of 20 to 30 l per kW heating capacity may be used.

Example:
Heating Capacity B0/W35: 8.9 kW
Volume of the Buffer Tank
= 8.9 kW x 20 l/kW = 178 l
A 200 litre buffer storage tank should be selected:

$$Q = m \cdot c \cdot \Delta T = P \cdot t$$

$$t = \frac{m \cdot c \cdot \Delta T}{P} = \frac{1000 kg \cdot 4187 Ws / kgK \cdot 5k}{5000 W}$$

$$= 4187 \ s = 1.16 \ hours$$

A 1000-litre buffer storage tank can fully meet a 5 kW heating demand for approximately one hour (at the full supply/return temperature difference).

9.4 Circulation pump

With the use of a parallel tank or a low loss header, a heating circulation pump and buffer circulation pump must be planned in, otherwise only a heating circulation pump is required.

The flow rate must be capable of transporting the full heating capacity of the heat pump.

The flow rate of the buffer feed pump and the heating circulation pump should be kept the same.

SIZING

The piping system and the circulation pump should be planned for a flow rate which results in a 5 K supply/return temperature difference in order to achieve the highest comfort level. A higher temperature difference (10 K) is acceptable for heating with radiators or individual radiator planning.

The resulting pump output required is therefore four times the required volume flow for a conventional system with a 20 K temperature difference.

$$\dot{m}_H = \frac{Pn \cdot 3600}{c \cdot \Delta t} \text{ (kg / h) Mass flow rate}$$

(Corresponds approx. to flow rate in l/h)
P_N = heating capacity = P thermal
c = Specific heat capacity of water = 4.2 $\dfrac{kJ}{kg \cdot K}$

(conversion factor 1 kWh = 3600 kJ)
Δt = temperature difference (for example, 5K)

The buffer circulation pump (if required) should provide the same flow rate. The pressure head is calculated from the head losses in the heat pump (see data sheet), fixtures and heat distribution system.

9.5 Connection group

Some manufacturers deliver pre-assembled pump/manifold units with a properly sized pump. These connection groups consist of buffer circulation and/or heating circulation pump, expansion vessel, ball valve, thermometer, manometer and insulation. In some models, the buffer or heating circulation pump is integrated in the heat pump.

The maximum flow velocity in the piping system should not exceed 0.8m/s (to minimize noise and head losses). The pipe cross-section should be sized accordingly (insulation at site).

9.6 Heating/cooling distribution system

The operation of the heat distribution system should be at the lowest possible temperature on grounds of:

■ Comfort: low-temperature radiation warmth provides optimal comfort

■ Health: lower air velocity, less dust turbulence

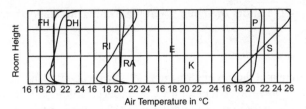

Characteristic air temperature profiles in center of room at steady state for different heating
systems with low outdoor temperatures.

FH = Radiant Floor Heating   K = Tiled Stove Heating
DH = Radiant Ceiling Heating   E = Iron Stove
RA = Radiator on outer wall under window
S = Gravity Air Heating with outlet on inner wall
RI = Radiator on inner wall   P = Perimeter Heating

Figure 9.3 Air temperature profiles
Source: Recknagel

■ Economy: the supply temperature determines the COP of the heat pump; lower
temperatures result also in lower losses in the system. With radiant warmth from
floor or wall heating, comfort is achieved with 2°C lower room temperatures than
with conventional radiator heating.

These requirements are optimally met with radiant floor/wall heating. Moreover, the
heating surfaces generally contain enough thermal mass to bridge utility shut-off
times.

Figure 9.3 shows the air temperature profile for different heating systems.

Radiant surface heating has an important self-regulating effect: the temperature
difference between the surface and room temperature is automatically small. If the
room temperature increases due to interior gains, the emission of warmth is
correspondingly reduced immediately.

9.6.1 Radiant floor heating

When used with heat pumps these systems should ideally be planned for
supply/return temperatures at 35/30°C. The installation of the radiant floor heating
system should be completed in accordance with manufacturer guidance. A standard
radiant floor heating system is only sensible in a well insulated building.

In Figures 9.5 and 9.6, a comparison of different floor constructions is shown.

Radiant heat is very effective at producing comfortable warmth. We feel as
comfortable in a room that is heated to 20°C by radiant heating as we do in a room
heated to 22°C by a conventional heating system. This is not only more comfortable
and healthier, its saves a tremendous amount of heating costs. Rule of thumb: a 2°C
reduction in room temperature saves up to 10 per cent of heating costs annually.

The combination of radiant floor and wall heating (heating piping placed in a
wall is perfect for radiant heating) is one of the best forms of heat distribution for
your energy-conserving house.

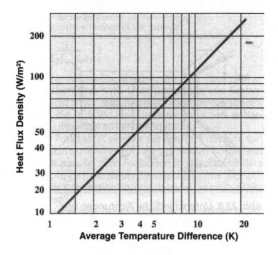

Figure 9.4 Characteristic line (only for isothermal surface)
Source: Udo Radke; Surface heating/surface cooling

INITIAL HEATING OF SCREED

The initial heating of anhydrite screed should be completed as instructed by the manufacturer. The initial heating of cement screed should occur after a minimum curing period of 21 days. Use caution if using an automatic heat-up programme on a heat pump (electronically controlled). The capacity of a ground heat source should be taken into account.

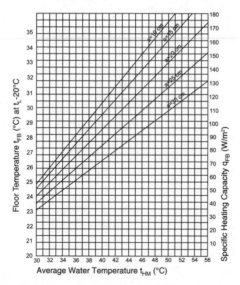

Figure 9.5 Performance diagram for surface R1 = 0,10 (m²K/W) carpet, 22mm thick parquet
Source: Ochsner

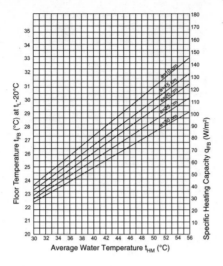

Figure 9.6 Performance diagram for top layer R1 = 0.02 (m²K/W) tile and stone surface, adhered PVC and linoleum up to 3mm
Source: Ochsner

9.6.2 Heating capacity and self-controlling effect
The characteristic line for the heat flux density shows the maximal heat transfer as a function of the temperature difference between surface temperature and room temperature.

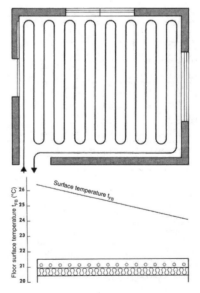

Figure 9.7 Switchback layout
Source: Ochsner

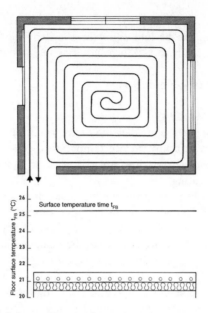

Figure 9.8 Snail layout (basic form, without border zone)
Source: Ochsner

An increase in temperature difference of the heating fluid temperature to room temperature at + 9 K by only 1 K leads to a heating capacity change of approximately 11 per cent.

If the room temperature increases from 20°C to 21°C when the surface temperature of the floor remains a constant 25°C, the heating capacity is reduced by about 20 per cent. (See Figure 9.4.)

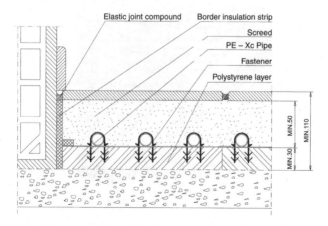

Figure 9.9 Construction example, secured with fasteners
Source: Ochsner

PIPING LAYOUT

The radiant floor piping can be laid either in a switchback (Figure 9.7) or snail pattern (Figure 9.8). Naturally, mixed forms are possible in border areas.

FLOOR CONSTRUCTION (see Figure 9.9)

9.6.3 Radiant wall heating

The size of the radiating surface enables the desired comfort level to be achieved, as with the tiled stove, with a low surface temperature. Normally 25°C to 28°C is enough. The mild radiant warmth does not stir up dust, as is often the case with radiators or forced-air heating.

Basically, radiant wall heating offers the same benefits as radiant floor heating. Because comfort is also dependent on the temperature of the walls, radiant wall heating offers additional comfort. With warmer walls, the body radiates less heat to its surroundings. This means that comfort can be achieved at lower air temperatures, which translates to lower heating costs and increased oxygen content in the air.

The heating capacity dependence upon pipe spacing and supply temperature is similar to that with the radiant floor heating. Because of the increased portion of convective heat transfer, this system is also well suited for radiant **cooling**. With a heat transfer coefficient of $\alpha = 8$ W/m²K, the heating and cooling capacities are similar.

There are certain preferred applications, for example small bathrooms, where the radiant floor heating is insufficient. In this case, the heating pipes are placed in both the floor and wall (Figure 9.10). In old buildings it is often the case that the existing floor must remain intact and the contractor does not want to be bothered

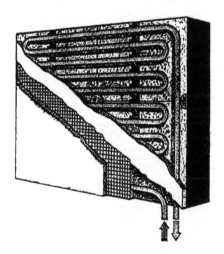

Figure 9.10 Cross-section of radiant wall heating system
Source: Bauratgeber

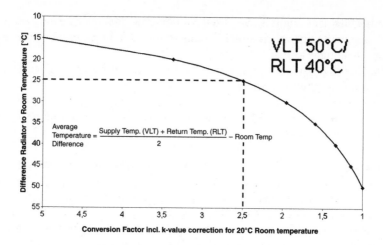

Figure 9.11 Size of radiator as a function of supply temperature
Source: DIN EN 442

by numerous radiators. Radiant wall heating is a problem free solution. Radiant floor and wall heating can be easily combined if desired.

The warmer supply is near the floor, the cooler return near the ceiling. The resulting effect is that more heat is distributed near the floor than at head level.

Another installation possibility in old buildings is to lay the heating pipes in grooves cut into the surface. There are also applications for finished dry-slab systems (ensure proper installation).

9.6.4 Radiators

The heating supply temperatures should be limited to 55°C (max 65°C for renovations) with heat pumps. For this reason, low temperature radiators should be planned. Figure 9.11 provides a guideline for the determination of the required heated surface area increase in comparison to conventional heating.

The combination of radiant floor and radiator heating is possible and in certain situations (wooden ceilings) sensible.

The installation of convection fans (fan-coils) can also be planned. Advantage: larger heating/cooling capacity of a room with a reversible heat pump (reversible system for cooling using the existing heat distribution system – see Chapter 10).

9.7 Domestic hot water production

To achieve maximum system efficiency, a separate, independent heat pump should be planned for domestic hot water heating (see Chapter 3.7). This should be optimally sized and installed and can provide additional functions (ventilation, cooling, dehumidifying). Units which make use of exhaust air as a heat source are referred to as exhaust air heat pumps.

The heating heat pump is over-sized for domestic hot water heating during the summer and must operate at a higher temperature difference in the winter due to the water heating priority.

If the heating system is to be used for domestic hot water production, notice should be taken of the following:

■ Heating demand of 1–1.5 kW (single family home) to be accounted for when sizing the heat pump.

■ Hot water tank for a three to five person household is approximately 300 l.

■ Plan the heat exchanger for a 5 K temperature difference with a supply temperature of 55°C or 65°C. This results in about 50°C or 60°C hot water temperature produced by the heat pump (if additional electric heating is used, the operational cost will increase and carbon emissions may also increase).

■ The heat exchanger must be capable of transferring the entire heat pump capacity (or at least the capacity of one compressor in dual compressor systems).

■ For internal heat exchangers, a surface area of 0.4 to 0.7 m² per kW heating capacity is recommended (for steel tubing). Where this cannot be achieved, then plate heat exchangers will be required. Small coils in DHW tanks will not work effectively with heat pumps.

■ A heating capacity of 0.25 kW per person for hot water heating should be accounted for when sizing the heat pump (average single family house and demand).

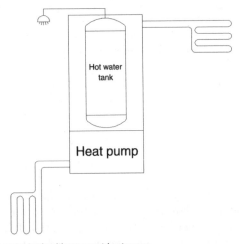

Figure 9.12 Integrated hot water tank with compact heat pump.
Source: Ochsner

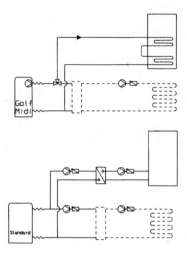

Figure 9.13 Basic schematics for domestic hot water heating with heating heat pump and hot water tank
Source: Ochsner

■ With Compact Heat Pumps the hot water tank may be built in.

Basic schematics for domestic hot water heating with heating heat pump and hot water tank:
a) Domestic hot water heating with Double Heat exchanger Tank
b) Domestic hot water heating with External Plate Heat Exchanger
c) Heating heat pump (compact unit) with integrated hot water tank

Any local regulations applying to hot water provision must be taken account of. In particular, the regulations relating to pressurized (unvented) hot water cylinders must be adhered to. Although heat pumps cannot cause boiling in hot water tanks, immersion heaters, solar panels and/or other secondary sources that are connected to the hot water tank can potentially cause boiling and associated pressure excursions.

10

Heating and Cooling – Reversible Heat Pump

10.1 Response to climate change

The reversible heat pump has been a standard design for all climates where air conditioning is essential. In 2003, Europe experienced the hottest summer in 200 years. Even in Austria and Germany there were over 49 days with temperatures above 30°C. The undeniable climate change offers an additional application for heat pumps in a heating/cooling configuration, even in moderate climates.

For little additional cost, a standard add-on is available for both ground source and air/water heat pumps, enabling cooling operation during the summer.

Cooling with an air/water or water/water heat pump offers considerable advantages when compared to traditional air conditioning systems: The heat pump can be less expensive in terms of both initial investment and operating costs. Additionally, they are quieter and healthier, as any draughts are avoided.

In offices and public buildings, a maximum room temperature must be provided. In Germany, for example, it is 26°C.

Environmental concern also presents a reason for cooling with efficient heat pumps: 'Green Cooling' can help significantly in lowering global CO_2 emissions.

10.2 Technical requirements for buildings

Due to the good insulation commonly found in buildings today, the required heat removal per square metre during summer for cooling is of the same order as the required heat input during winter in central Europe (with the exception of buildings with large window areas exposed to direct solar radiation or hot climates).

With the now common low surface temperature heating systems, rooms can be heated with a low heating supply temperature (35/30°C), resulting in a small temperature difference between the heating surface (floor, wall) and the room temperature, efficiently and comfortably heating the room.

As is the case with heating, the heat flux density in cooling is dependent upon surface temperature and room temperature. The following characteristic line allows for a qualitative analysis:

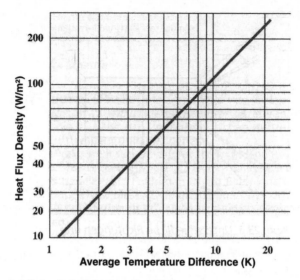

Figure 10.1 Characteristic line (only for isothermal surface)
Source: Udo Radke; ABC's of Surface Heating/Surface Cooling, 2005

With 16 to 18°C supply temperatures and small pipe spacing, surface temperatures (floor, wall) between 19 and 21°C can be achieved, maintaining room temperature between 22 and 26°C with outdoor temperatures of 32°C. (Caution: solar influences and large window areas must be accounted for.)

Consequently, the required mass flow rates for radiant wall heating are also sufficient to meet cooling demands.

In combination with fan coils or domestic hot water heating via an exhaust air heat pump, the air can also be dehumidified.

10.3 Planning for active cooling

The required temperature level would be available directly with ground source collectors, but only in the short term. The problem is that with this passive 'free cooling' process, the temperature of the heat sink (earth) rises quickly. After only a few consecutively hot days, the cooling effect may be negligible. Additionally, limitations on mass flow rates can reduce the effectiveness of such systems. The 'activation' of structural building elements can have positive effects, but it does not provide sufficient cooling capacity.

With active cooling the refrigeration cycle of the heat pump is reversed using a four-way valve. The condenser is transformed into an evaporator and actively removes the unwanted heat (cooling) from the floor/wall areas and places it into the ground.

COOLING CAPACITY
The cooling capacity during cooling operation is often higher than heating requirement during heating operation.

HYDRAULICS
The expected temperature difference should be considered when sizing the heating/cooling system circulation pump.

10.4 Sizing of ground collectors and vertical loops

With active cooling (reversal of refrigeration cycle), sufficient heat can be transferred even with elevated ground temperatures. In central Europe, the heat input into the ground during the summer is negligible. For this reason, seasonal cooling operations require no special permits. The collector/loop must be in accordance with relevant standards (i.e. VDI 4640 or ÖVGW). Sizing will be similar to that for heating purposes.

The heat input also helps to fully regenerate the earth for the heating season. If the system is to be operated year-round, the collector system must be sized accordingly.

Figure 10-2 Convection fan
Source: Ochsner

10.5 Cooling distribution system, radiant/surface/panel cooling

The cooling distribution system consists of a radiant wall/floor heating system with narrow pipe spacing, planned for a maximum supply temperature of 35°C. With a radiant floor heating system, adding an additional wall area (40 per cent) is recommended. This ensures healthy and comfortable silent cooling without draughts or noise. If necessary (i.e. large window area), the installation of convection fans is recommended. With these fan-coils, a large cooling capacity can be comfortably provided. These devices are usually equipped with multi-level fans, allowing the desired cooling performance to be easily controlled.

The hydraulic system is generally identical to that used with a heating system. This eliminates the need for any additional piping for active cooling. With many heating/cooling units, domestic hot water heating is available optionally.

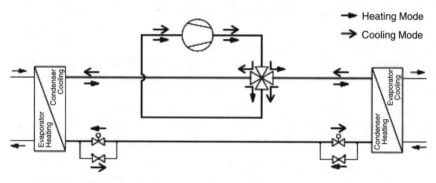

Figure 10.3 Reversible refrigeration circuit for a brine/water heat pump
Source: Ochsner

10.6 Refrigeration circuit for active cooling

Reversible heat pumps, are capable of heating or cooling via a reversible refrigeration circuit: The compressor still pumps in the same direction, but a four-way valve is used to switch the heating condenser into an evaporator and the heating evaporator into a condenser.

The heat source and heat sink are 'swapped'.

Heat pumps using outside air as the heat source are usually already equipped with a reversing device, since this is needed to provide defrosting capability (de-icing of the finned evaporator).

A refrigeration circuit suitable for reversible operation can be implemented at different levels of complexity, which in turn achieve different coefficients of performance in reversible operation.

10.7 Control

The control unit for heating/cooling often maintains fully-automatic load conformity (cooling curve/sliding control). The minimum supply temperature is set such that build-up of condensation is avoided. An additional sensor is also available, which provides instantaneous dew-point values, ensuring the avoidance of condensation. The conversion from heating to cooling operation is automatic.

10.8 Economics

The efficiency, EER, of the climate control heat pump describes its cooling efficiency. The operating costs of a heat pump in cooling operation depend on the COP actually achieved:

$$\varepsilon = COP = \frac{\text{Delivered heat energy (kW)}}{\text{electrical input to compressor (kW)}} = \frac{\text{Environmental energy} + \text{el.i.to.c.}}{\text{el.i.to.c.}}$$

el.i.to.c. = electrical input to compressor.

During heating operation, the power required by the compressor is usable warmth, whereas during cooling operation this is considered a loss which must be removed.

The energy efficiency, as well as the operation costs of an air conditioning system or a heat pump in cooling operation, depends mainly on the required temperature difference. The **EER** is the corresponding **performance coefficient**:

EER = Cooling Capacity/Required Electrical Power
i.e. = 3kW Cooling Capacity/1kW Required Electrical Power = EER3

In practice, ground source climate heat pumps require only a small temperature difference. In long-term operation, the collector area reaches a maximum of 25°C. The temperature difference between the cooling supply (15°C) and the brine ground collector (25°C) is only 10 K, which allows for a high COP and EER.

Example: GMSW 15 H/K (optimized for cooling operation)
Operation point B25/W18 (Long-term operation)

Cooling capacity	17.7 kW
Power consumption compressor	2.9 kW
Power consumption brine pump	0.3 kW
Power consumption storage tank pump	0.09 kW
Total power consumption	3.29 kW
EER (including pumps)	5.4

Conclusion: The operation costs of a Climate Control Heat Pump/Ground Source System is less than half of that of a conventional air conditioning system – in this example.

10.9 Solar cooling

In the attempt to utilize alternative energy sources for cooling, solar-driven systems including both absorption (fluid substances) and adsorption (solid substances) refrigeration machines have been investigated.

With high investment costs and COP values from 0.3 to 0.8, solar cooling has very few practical applications. The most economical systems are ground source systems, such as brine/water ('Green Cooling').

If properly sized, even air/water heat pumps can offer cheaper operation than traditional air conditioning systems (in addition to an increased comfort level).

10.10 Groundwater heat source

Water/water systems are the only systems in which reversible refrigeration operation is theoretically unnecessary, and then only if a specialized hydraulic system is in place. A two-stage operation, first passive, then active, is also possible.

10.11 Operating range

With respect to heat rejection into the earth, there is no practical limit for cooling operation in seasonal operation. With regard to using the standard heating system as the cooling distribution system, operation is limited by possible condensate build-up, the result of the supply temperature falling below the dew-point temperature. In central Europe, this is generally not a problem with a supply temperature of 17°C (17/22). Local conditions and climate data should be taken into account. If necessary, a condensate collector should be installed with fan-coils. Installation of cooling loops with supply temperatures as low as 12°C is feasible (12/17), but needs special design and AC-experience.

11

Control of Heat Pump Heating Systems, Electrical Connections

11.1 Controller and control system

The purpose of the controller is to hold the temperature of a given room at a nearly constant, preset temperature. This is achieved by controlling the heating supply temperature (control variable). Either the room temperature or the outdoor temperature is used for the reference variable. The simplest form of control is achieved with a simple indoor thermostat.

A *weather-driven* controller offers considerably more comfortable heating. These systems include an outdoor sensor which enables the controller to maintain a nearly constant room temperature, even with extreme changes in outdoor temperature. The weather-driven controller uses a programmed heating curve to account for building characteristics and the properties of the heating system at varying outdoor temperatures. The heating curve is a straight line which is defined by two points, and is set specifically for each system (via y-intercept and slope). With additional room thermostats, the influence of other heat sources can be considered.

In systems without a low loss header, the hot water sensor is placed in the return, allowing the controller to account for conditions in the heating distribution system. In systems with a low loss header, the sensor is located within the separator (adhere to manufacturer's guidance).

11.2 Heat pump control

The heat distribution control is achieved in the same manner as with an electric heater or boiler, by switching the heater on/off (two-point control). A properly planned heat distribution system is designed to release heat in a manner that reduces large temperature swings in living spaces.

In large heat pump systems, stepped control is common (modular construction, multiple compressors etc.). In such systems, the auxiliary systems (fans, circulation pumps etc.) should operate in a similar manner in order to maintain the highest possible coefficients of performance. A further possibility is the use of a frequency converter to allow for variable drive control. In this case, the proper lubrication of the compressor must be ensured. Furthermore, the losses associated with the frequency converter must be taken into consideration.

11.3 Additional functions of heat pump control

Depending on the manufacturer's configuration, aside from the control of supply and return temperatures, the controller must complete other tasks:

- limit start/stop frequency (start delay)

- defrost of the evaporator with air/water heat pumps

- regulate operation to match discounted electric rates

- optional domestic hot water heating

- optional mixing control

- additional heat source control etc.

- safety functions.

11.4 Hydraulic schematics

The hydraulic schematics intended for heating operation are included in the chapter 'Heating Systems'. It is important that the heating loop is operated such that the heating supply temperature is the temperature produced by the heat pump, except if a mixing valve is absolutely necessary (combined radiant floor/radiator systems).

11.5 Single room control/zone control

Single room control is commonly achieved with the installation of thermostatic control valves for individual heating surfaces or zones. Centralized control is possible using a central building control/BUS.

Systems with hydraulic separation are recommended to improve operation reliability.

11.6 Other systems

Not every heat pump system uses a hydronic heat distribution system. For systems with direct condensation, room or zone thermostats are used. The same is true for air/air heat pumps.

11.7 Electrical connections

Electrical energy is sometimes available at special rates (heat pump tariff) designated for heat pump operation.

This may be an interruptible service, or may only be available at certain times of day or night (see local utility guidelines).

MEASUREMENT AND CONTROL EQUIPMENT

In some countries, when using special electric rates, the acquisition and installation of up to two additional electricity metering devices may be required:

- one for metering consumption at the special rate

- one for a rate control device (pulse control receiver or timer)

The switching from high to low rates or vice versa is controlled with these additional meters. The switching device is usually provided and installed by the utility in the space provided.

ELECTRICAL CIRCUIT SEPARATION

For bivalent and interruptible monovalent operation heat pumps, it must be possible to temporarily interrupt the power supply. For this reason, two separate electrical circuits are required:

- The 'controlled main electric circuit' (heat pump meter) is for the compressor and the drive power for the heat collector system (fan, brine or groundwater pump, all electric heating elements). A lead circuit switch should be mounted in the upper connection space above the heat pump meter for regulation of the electrical circuit.

- The 'uncontrolled electric circuit' (household meter) is for all connections which must operate without interruption:

    - Control of the complete heating system

    - Circulation pump for the heating and hot water systems

CONNECTION (see Figure 11.1)

The heat pump electrical connection may only be completed by a qualified electrician. The guidelines of the local utility must be adhered to. Where it is required, the connection proposal to the utility is to be completed by the electrician.

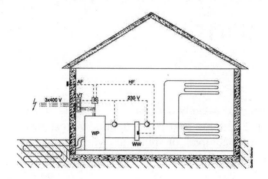

Figure 11.1 Electric connection sketch
Source: Ochsner

12

Domestic Hot Water Production

Hot water heat pumps are specially designed for domestic hot water preparation (domestic hot water heat pumps, heat pump boilers and exhaust air heat pumps). They are offered as either compact or split units. Various units also offer the user free side benefits (multi-function units).

12.1 General instructions for domestic hot water

12.1.1 Why not with the heating loop?

As oil-fired central heating systems became standard in the 1960s, the comfortable understanding was that the systems could be used for domestic hot water heating year round.

From today's point of view, it is a brutally false conclusion to consider domestic hot water as a 'cheap' byproduct of central heating. This is particularly true for oil and gas boilers outside the heating season, as these units are designed to heat the building and are grossly over-sized for hot water production only.

12.1.2 Domestic hot water system comparison

By the time the burner, boiler, flue and heat exchanger have reached operating temperature, it is often the case that just as much energy has been expended as is actually required for the water heating itself.

These high 'startup losses' are intolerable when ever-increasing energy prices are considered, not to mention the disproportionate environmental impact these systems cause.

When heating domestic hot water in summer months with an oil-boiler, as little as 25 per cent of the primary energy consumed is converted into energy in the form of hot water.

Heating heat pumps must operate with a higher temperature difference for domestic hot water heating than for normal heating operation, reducing the coefficient of performance. For this reason, the installation of a separate domestic hot water heat pump is logical.

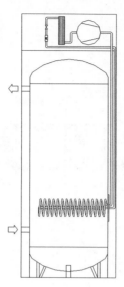

Figure 12.1 Hot Water Heat Pump
Source: Ochsner

OPERATION COSTS AND ENVIRONMENTAL IMPACT COMPARISON FOR DOMESTIC HOT WATER SYSTEMS

HEAT PUMP

A COP = 3.5 represents a system efficiency of 350 per cent of the electrical energy consumed, as 250 per cent cost-free environmental energy is used.

OIL BOILER (see Figure 12.2)

System Efficiency from 14% to 40%

$$\eta_{sys} = 100 - q_a - q_s - q_b - q_v$$

1   Exhaust losses q_a = 10* to 25**%
2   Radiation losses q_s = 1 to 5%
3   Startup losses q_b = 48 to 54***%
4   Distribution losses q_v = 1 to 2%
    System losses Σ = 60 to 86%

(Partial efficiency according to VDI 2067)
*New system   ** Old system   *** Summer operation

Figure 12.2 shows a schematic representation of hot water preparation via a heating boiler.

SOLAR SYSTEMS (see Figure 12.3)
The solar fraction of a simple standard system is only 68 per cent (according to ASSA, Prof. G. Fanninger, circulation Pump 70 W x 6 h/d). Larger collector areas, a

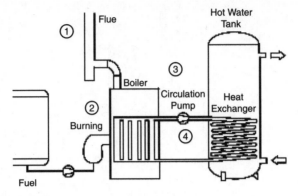

Figure 12.2 Heating Boiler
Source: Ochsner

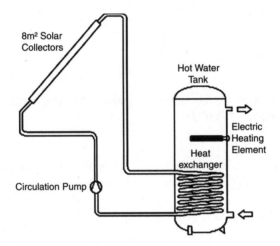

Figure 12.3 Solar system
Source: Ochsner

larger storage tank and vacuum tube collectors can all increase the solar fraction, but at significant increases in investment costs. Generally, the investment costs for a solar system greatly exceed those for a heat pump system. When considering a solar system, operating expenses for circulation pumps and auxiliary heating during poor weather periods must be accounted for.

Electric boiler (see Figure 12.4)

COP = 1 represents a system efficiency of 100 per cent of the electrical energy consumed.

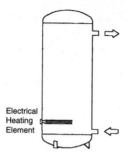

Fig. 12.4 Electric boiler
Source: Ochsner

Cost comparison in Figure 12.5 is an example for one specific set of tariffs

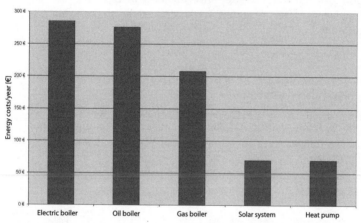

Fig. 12.5 Cost comparison for various methods of domestic hot water preparation
Source: Ochsner, based on VDI-data

12.1.3 Separation of domestic hot water from the heating system

This is important for the following reasons:

OIL + GAS HEATING

- high startup losses

- poor system efficiency, particularly in summer

- disproportionally high emissions

■ danger of soot build-up in chimney

■ important for units taking advantage of the higher heating value (condensing boiler units).

HEATING HEAT PUMP

■ For domestic hot water heating (boiler priority), supply temperatures of 55°C are necessary, decreasing the COP from the value that could be achieved when used exclusively for a low-temperature heating system. For this reason, a separate domestic hot water heat pump is the optimal solution.

■ Domestic hot water heat pumps offer useful, cost-free byproduct functions such as ventilation, cooling, dehumidification etc.

■ For domestic hot water heating with a heating heat pump, sufficient sizing of the heat exchanger (see Chapter 4.4.3) as well as the system are critical (see Chapter 9.7).

12.1.4 Information for planning and operation of a domestic hot water system

The maximum operating temperature in the distribution network is 60°C according to the Energy Conservation Act (Energieeinsparverordnung, Germany).

Operating temperature for domestic hot water heat pump 55°C, size storage tank accordingly.

Avoid secondary hot water circulation ('hot loops'); as a minimum, use a timer and ensure that the pipework is well insulated.

12.1.5 Installation instructions

Table 12.1 Piping specifications

Outer diameter mm	Nominal diameter mm	Volume (l/m)	Heat dissipation from 1m insulated piping for hot water temperature 60°C and surrounding temperature 20°C	
			W/m	kWh/m.a.
18	15	0.20	6.3	55
22	20	0.31	6.8	60
28	25	0.49	7.2	63

Source: VDEW

■ Do not use copper tubing before steel (in the direction of flow).

■ Insulate piping

■ Minimum insulation thickness (with $\lambda = 0.035$ W/mK)
up to DN 20: 20mm insulation
up to DN 35: 30mm insulation.

12.1.6 Hot water demand – guidelines

Table 12.2 Household

Water quantity	at 40°C	at 50°C
Full bath	120–150l	96–120l
Shower	30–50l	24–40l
Hand washing	2–5l	1.5–4l

Source: VDEW

(The average domestic hot water demand is 30 litres/person/day at 45°C.)

Table 12.3 Commercial

Water quantity	at 50°C
Inn	Litre/guest/day
Room with shower	70
Room with bed	140
Hair salon	Litre/guest/day
Men	80
Women	160

Source: VDEW

12.1.7 DVGW – worksheet W551

This worksheet deals with the avoidance of dangerous Legionella build-up in drinking water installations through which heated water is transported. Systems are differentiated as small systems (i.e. single/double family homes) and systems with over 400 litres tank volume and over 3 litres piping volume (without circulation), and large systems. For large systems, operation temperatures must reach 60°C through a daily heating treatment. This heating process may be completed using the appropriate heat pump or an electric heating element set on a timer.

12.2 Additional optional functions of hot water heat pumps

Hot water heat pumps can offer additional benefits besides hot water production. With the powerful radial fan and the optimally designed spiral housing provided by some manufacturers, air ducts over 20 metres in length can be connected to serve as ventilation.

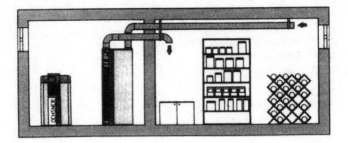

Figure 12.6 Intensive cooling (air recirculation operation)
Source: Ochsner

12.2.1 Additional cooling benefit

In the variation depicted in Figure 12.6, the inlet and outlet are placed in the room to be cooled (store room, wine cellar, etc.)

In the variation depicted in Figure 12.7, only the outlet is placed in the room to be cooled. An exhaust air opening or open window is necessary: store room, bedroom, lounge etc.

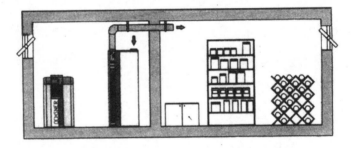

Figure 12.7 Lighter cooling effect (fresh air operation)
Source: Ochsner

12.2.2 Additional basement dehumidifying benefit

In order to make optimal use of the basement dehumidifying function, the inlet and outlet should both be located in the room to be dehumidified (recirculation operation). The heat pump then operates as a dehumidifying device by removing the condensation water (see Figures 12.8 and 12.9).

A limited cooling or dehumidifying effect is achieved with fresh air operation, which is comparable to using a fan to ensure fresh air exchange.

12.2.3 Additional distilled water benefit

The condensation water is mineral free and can be used in humidifiers, steam irons etc., but should be considered non-potable.

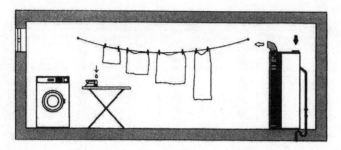

Figure 12.8 Installation for drying function in laundry room (recirculation operation)
Source: Ochsner

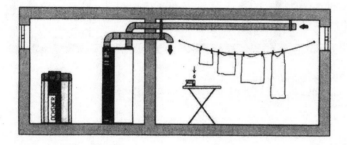

Figure 12.9 Drying function in neighbouring basement room with air duct (recirculation operation)
Source: Ochsner

12.2.4 Additional ventilation benefit

The following ventilation functions are possible:

SIMPLE EXHAUST-AIR VENTILATION FUNCTION (SINGLE DUCT SYSTEM)

■ Suitable for year-round operation.

■ The air intakes are located in damp rooms. The heat pump also uses the warmth from these rooms.

■ The processed exhaust air is then to be vented outdoors.

■ It should be confirmed that fresh air can enter the individual room (fresh air vents, door grilles; see diagram).

■ Prerequisites for the use of a hot water heat pump as a ventilation unit includes sufficient fan performance (radial fan, spiral housing), as well as ventilation operation independent of hot water preparation. If these requirements are met, the unit can be considered a multi-function unit.

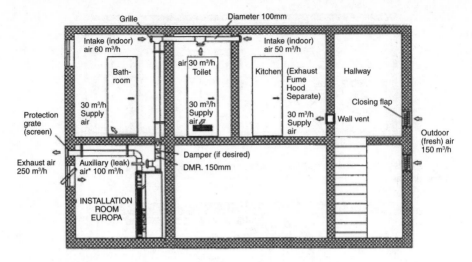

Figure 12.10 Example: multi-function unit Europa 312
* Auxiliary (leak) air necessary if air flow from damp rooms is reduced
Source: Ochsner

■ Room cooling via fresh air supply (summer).

■ With this simple ventilation system without fresh air heating, it is recommended to reduce the air intake from damp rooms and increase air intake in the heat pump installation room during periods of very low outdoor temperatures (easily achieved with an adjustable damper). The ventilation system can be completely suppressed with this method.

SIMPLE EXHAUST-AIR FUNCTION FOR YEAR-ROUND OPERATION
Figure 12.10 depicts heat recovery using a heat pump for hot water preparation (schematic view):

■ Sound insulation is recommended on ducts.

■ With interconnecting rooms, ensure connections/ducts properly sized.

■ When using an exhaust air heat pump in buildings with an open fireplace, the air intake dampers should be sized for a maximum 4 Pa negative pressure.

12.3 Installation

■ Any dry, frost protected room is suitable for installation.

■ It is important that the volume of the room is sufficient.

■ The intake temperature should not fall below the heat pump operating range. For units with defrost functions, this range is practically unlimited. The distance to the intake point should be kept as short as possible.

CLEARANCE
A clearance of 20cm to walls should be allowed. The location is to be selected such that operation and maintenance are possible.

12.4 Water connections

12.4.1 Water connections
The hot water tank connections must adhere to local standards. For technical data, see detailed descriptions provided by the manufacturer.

12.4.2 Air connections
The air duct connections are oriented differently depending on the manufacturer. The most convenient placement is on the top of the unit, with 150mm or 160mm diameter sleeves. With some manufactures, it is possible to change the placement of the connections to the back of the unit simply by rotating a cover panel.

12.4.3 Air ducts
For air supplies from another room, ensure that the volume of the room is sufficient. The air ducts should be smooth and have a diameter of at least 150mm. The maximum length is 20m with a maximum of three 90° bends. Caution: with units with axial fans, only short (if any) ducts can be used.

12.4.4 Condensate water drain
During heat pump operation, up to 0.3l/h condensate can accumulate. The condensate drain is frequently located on the backside of the heat pump.

The condensation drain should not be connected directly to drain lines. A trapped connection to the drain line is to be used.

12.4.5 Ground collector connection
Heat pumps, which do not use energy from the exhaust air, but rather from the ground, must be properly connected to the ground collector (generally pre-filled with quick-connections). A proper wall break-through must be ensured.

12.4.6 Heating heat exchanger
Models fitted with a heating heat exchanger can also be interconnected to an existing heating system. The supply and return of the heat exchanger are to be connected to the heating system and a circulation pump with check-valve should be installed. Figures 12.11 and 12.12 show connection schemes for domestic hot water heat pumps

with hot water priority, with an oil or gas boiler or with a solid fuel or pellets oven. Solar connections are also possible.

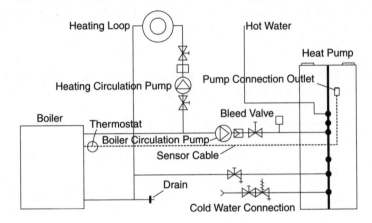

Figure 12.11 Connection schematic for domestic hot water heat pump with oil/gas boiler and hot water priority
Source: Ochsner

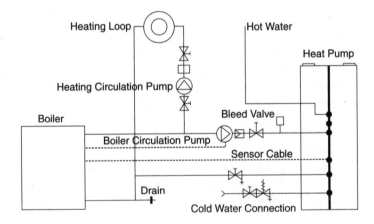

Figure 12.12 Connection schematic for domestic hot water heat pump with solid fuel or pellet oven and hot water priority
Source: Ochsner

13

Controlled Dwelling Ventilation

13.1 Why ventilate?

As early as 100 years ago, Father Kneipp said: 'The same air breathed three times has toxic effects. How much discontentment and sickness people could save themselves with fresh, oxygen-rich air...' (Source: The Kneipp Health Book).

Today, people understand the importance of *indoor air quality* for *good health*. Good air quality also influences people's perception of *comfort* considerably. Moreover, air quality influences the preservation of *building materials* and *stored goods*.

According to diverse studies, housewives spend 87 per cent and working women 56 per cent of their day in the home. These times of occupation show that at least the minimum standards for room climate should be met. Room climate includes

Fig. 13.1
Source: Energie AG

temperature (room, wall), humidity, air mixture, air speed and other physical air conditions. Indoor air quality includes air mixture, composition and impurities; in other words, the content of so-called pollutants. In addition, noise from outside is always to be considered.

CAUSES OF POOR AIR QUALITY

Building materials, emissions from people (water, carbon dioxide, organic substances), tobacco smoke, mould and mites, pollen, food, stored goods, radon, water vapour production while cooking, washing, bathing, byproducts of combustion, chemicals for household cleaning (soaps, detergents, sprays, cosmetics), evaporation of solvents and adhesives, paints, lacquers, furniture, rugs and textiles, dust from clothing and furnishings, and house pets.

These pollutants concentrate and can only be removed from living spaces with sufficient exchange of fresh, oxygen-rich air. In Table 13.1, moisture emissions from various sources in gram per hour are shown.

Table 13.1 Moisture emissions

Source		Moisture emission
Kitchen	Cooking	600–1000
House plants	Average size	5–20
Clothing dryer	Dripping wet	100–500
	Spun dry	50–200
Bath	(each)	700
Shower	(each)	2500
People	Depending on activity level and clothing	30–3000

Source: VDI

13.2 Why controlled dwelling ventilation?

The *airtight* building materials, including windows and doors, which are used today in order to save energy, prevent the natural exchange of sufficient fresh air in living spaces.

THE EFFECTS

■ health damaging growth of mould

■ high concentration of microorganisms and mites leading to allergies

■ headaches due to high CO_2 concentrations and lack of oxygen

■ excessive relative humidity with moisture on structural building elements.

Ventilation through windows means heating energy losses and compromising comfort. Slightly opened windows wastes heating energy. Cross-ventilation results in drafts and can be considered 'coincidental ventilation'. Street noise enters the home unhindered. Dust and soot enter the living space. For these reasons, *controlled dwelling ventilation* is the method which fulfils today's demands.

The base *air change rate* is the minimum ventilation rate which ensures the desired air quality. In average inhabited living spaces a factor of 0.5 is recommended, meaning that in two hours the volume of air in the living space should be replaced. The fraction of ventilation heating demand compared to the total energy demand has increased with the new building standards.

The percentage of ventilation heating demand out of the total energy demand of a single family home:

past ca. 20 per cent, today ca. 26 per cent, low energy house ca. 40 per cent.

13.3 Types of controlled dwelling ventilation

■ single duct system (exhaust system)/two duct system (supply and exhaust system)

■ central, decentralized

■ with/without heat recovery

■ heat recovery via heat exchanger and/or heat pump

■ with/without auxiliary heating function

■ with/without cooling operation.

Fig. 13.2 Controlled dwelling ventilation with kwl-centre
Source: Helios

13.4 Comprehensive living climate

There is more than just heating and ventilation when it comes to providing an optimal living climate.

This includes:

■ Low-temperature hydronic radiant heating (radiant floor/wall heating) heat pump systems with proper control units which make use of solar energy year round.

■ When necessary, 'climate control heat pumps' for cooling via the existing heat distribution system in summer.

■ Controlled dwelling ventilation with heat recovery and warming of fresh air with an exhaust/supply heat pump in winter.

■ Optional combined domestic hot water heating or cooling function in summer.

SYSTEM VARIATIONS
Example: Controlled dwelling ventilation with 3 kW heating capacity* and domestic hot water preparation.

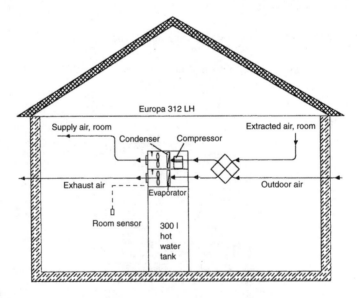

Figure 13.3
Source: Ochsner

* This represents the heating demand due to ventilation.

13.5 Combined units for heating, hot water preparation and ventilation

Numerous compact units which combine heating, hot water preparation and even controlled ventilation are on the market.

In some cases, combination storage tanks are used in these applications. In these tanks, the domestic water and heating loop water are stored separately. The heat sources range from outdoor air to exhaust air to ground or groundwater. In many cases, various heat sources are combined.

With the use of a conventional combination storage tank, it should be noted that it is less efficient (and more costly) to heat the domestic hot water via the heating water loop. This is because the heat pump is forced to operate with a higher temperature difference. There are, however, constructions (see Figure 13.4) which avoid these disadvantages, heating domestic hot water and the building with the highest coefficients of performance.

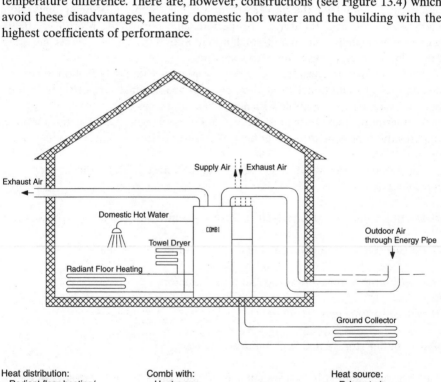

Heat distribution:
– Radiant floor heating/
  radiator heating
– Domestic hot water
– Supply air heating

Combi with:
– Heat pump
  combi tank
– Air/water heat pump (hot water)
– Ground source heat pump
  (heating system)
– Ventilation module
  (with cross-flow heat exchange)

Heat source:
– Exhaust air
– Outdoor air
– Electrical heating
  element (stand-by)
– Geothermal heat

13.4 Combi heat pump as central heater for heating, hot water preparation and controlled dwelling ventilation.
Source: Ochsner

14

Specialized Installations

14.1 Renovations

The stock of heating systems in single- and multi-family houses represents an important potential for the implementation of heat pump heating systems: after many years of use, buildings are periodically renovated. Moreover, many homeowners install additional thermal insulation to reduce heating costs and bring the building envelope up to modern standards.

This is the time to consider a new heating system. With the reduction in heating demand due to additional insulation, parts of the original and often over-sized heating system are compatible with a new heat pump system.

Furthermore, the German Energy Conservation Act (EnEV), for instance, requires the replacement of older heating systems before 31.12.2006 (boilers built before 1978).

For the assessment of heat pump compatibility, the following steps are recommended:

■ Heating demand and temperature readings before new insulation is installed.

■ Calculation of heating demand and temperatures after insulation is installed.

■ Check temperature difference and pipe diameters.

■ If necessary, increase radiator size.

■ Size buffer storage and heat pump accordingly.

The existing heat distribution system may in many cases be easily reconfigured with larger radiators, convection fan, dry floor or wall heating system.

A combination unit for controlled dwelling ventilation with heat recovery and air heating/cooling is also worthy of recommendation. Air heat source systems can be implemented almost anywhere. Air/water heat pumps are especially well-suited for renovation installations in bivalent operation (see Figure 14.1). With the use of a split air source heat pump, no modifications to the building structure are normally necessary.

The existing boiler may remain in place or can be replaced by adding an electric heating element to the buffer storage tank.

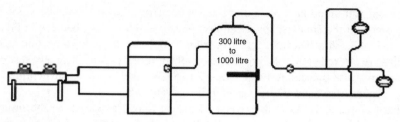

Figure 14.1 Air/Water Split Heat Pump System
Source: Ochsner

Bivalent-parallel, mono-energetic operation is ideal for renovations and can be implemented anywhere.

14.2 Low energy/passive house

The annual heating demand (heat quantity, energy demand) for low energy houses is about 60 kWh/m^2 per year. With about 1700 annual heating hours, this corresponds to a specific heating demand of approximately 35 W/m^2. A single family home with 130m^2 required a heat source with 4.5 kW heating capacity. A so-called 3-litre house with 130m^2 with approximately 15 W/m^2 specific demand requires a source with only 1950 Watt capacity. Conventional oil boilers or pellets are not available in capacities under 15 kW, and are therefore not practical for these applications. Due to the short operation period, gas connections are also relatively expensive. Solar systems are expensive and need stand-by electricity for bad weather. As a result, a heating heat pump system is ideally suited for the heating of low energy houses.

Figure 14.2 Heat pump Europa Mini EW for heat source ground with 4.8kW heating capacity and integrated loading pump
Source: Ochsner

Special mini-module heat pumps have been developed particularly for low capacities and the smallest space requirements possible, as can be seen in Figures 14.2 and 14.3. With heating capacities from 2 to 5 kW, these mini ground source systems are offered with brine ground collectors or direct expansion as a heat source. For systems with the smallest capacities (2 to 5 kW), the simple construction of the direct expansion system offers considerable advantages.

Space demand: These units have the smallest dimensions and can be installed without problem with brackets onto a wall. Despite the extremely quiet operation (scroll-compressors), the installation location should not be directly next to a bedroom.

Any preferred heat delivery system can be operated:

- radiant floor heating

- radiant wall heating

- radiator heating

- convection fans

- heating heat exchangers in air supply duct

- multi-purpose tank (solar combination)

- domestic hot water heater.

Additionally, low energy houses require controlled dwelling ventilation because of the tightness of construction. A comprehensive dwelling climate system can be ensured through the use of radiant surface heating.

Figure 14.3 Heat pump Europa Mini for heat source exhaust air with 2.2 kW heating capacity and integrated loading pump
Source: Ochsner

14.3 Swimming pool heating

The heating demand for an outdoor swimming pool depends upon the in-ground insulation, the location, and the profile of use. The heating demand can be approximated as follows for moderate climate:

with cover	50–150 W/m²
protected location	50–200 W/m²
partially-protected location	100–300 W/m²
unprotected location	200–500 W/m²

If appropriately planned, the swimming pool can also be heated by the heating heat pump. For ground source systems, the collector must be sized accordingly. The connection for swimming pool heating should be in parallel to the heating and hot water circulator pumps. The heating of the swimming pool water occurs with a swimming pool heat exchanger.

14.4 Livestock stable heat pump

Heat source system: in order to make use of the heat rejected by cattle, a silent evaporator in the form of a pipe heat exchanger is installed on a suitable stall wall (as shown in Figure 14.4). The pipes are corrosion resistant and remain undamaged by the dust and ammonia content in the air.

The heat pump system consists of three parts:

– heat source system
– heating heat pump
– heat distribution system/radiant floor heating.

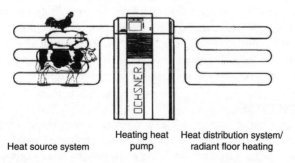

| | Heating heat | Heat distribution system/ |
| Heat source system | pump | radiant floor heating |

Figure 14.4 Stable Heat Pump
Source: Ochsner

Table 14.1 Heat rejection by animals

Dairy cows, breeding calves, breeding bulls, young breeding cattle
(Optimal stable temperature 16°C, relative humidity 80 per cent)

Animal Weight (kg)	100	200	300	400	500	600	700	800
Heat Production/Animal (W)	261	452	621	766	887	986	1050	1114
Heat Production/Animal (kcal/h)	225	390	535	660	765	850	905	960

Young sows, expectant and non-expectant sows, boars
(Optimal stall temperature 12°C, relative humidity 80 per cent)

Animal Weight (kg)	40	60	80	100	150	200	250	300
Heat Production/Animal (W)	104	139	168	197	269	341	414	487
Heat Production/Animal (kcal/h)	90	120	145	170	232	294	357	420

Source: Ochsner

14.5 Absorber

The so-called massive absorber is a special form of heat source system. They consist of large construction or building elements (roofs or walls), which contain piping for the collection of heat. Common are 'energy-fences', energy roofs (on garages, for example), as well as large paved areas.

Absorbers are also used with activation of building elements (passive, without heat pump).

14.6 Heat pumps for industrial applications

APPLICATIONS

Large capacity heat pumps are used in the heating/cooling of manufacturing facilities, office buildings, housing developments, administrative buildings, hotels and recreational facilities, sometimes making use of industrial process waste heat.

The resulting energy and operation cost savings, when compared to conventional heating and cooling systems, make the implementation of heat pumps particularly economical. Once units reach a certain capacity, they are ideally suited for use of low temperature district heating or rejected heat from other sources, such as industrial processes, exhaust air, geothermal energy and others.

REFRIGERATION CYCLE

Large heat pumps are equipped with semi-hermetic compressors. Depending on the system capacity, the compressor(s) is/are either multiple cylinder reciprocating compressors, screw compressors or turbo compressors, which operate alone or in parallel. Screw and turbo compressors contain no reciprocating parts and therefore operate with little vibration and noise.

Figure 14.5 Brine/water heat pump with 270 kW heating capacity for the heating of a sporting facility. The heat source consists of vertical ground loops with a total length of 3000 metres
Source: Ochsner

Figure 14.6 Water/water heat pump with screw compressor and shell and tube heat exchanger
Source: Ochsner

Figure 14.7 System with screw compressor model SVW
Source: Ochsner

For the evaporator and condenser, tube and shell or flat plate heat exchangers are generally used for brine/water (Figure 14.5) or water/water (Figure 14.6).

14.7 Industrial waste heat use/indirect water source

Exothermic chemical processes or various manufacturing processes require cooling. In many cases, heat pumps can be used to remove heat from the cooling water. This provides additional cooling for the process as well. Intermediate heat exchangers should also be used with conterminated or cold ground water.

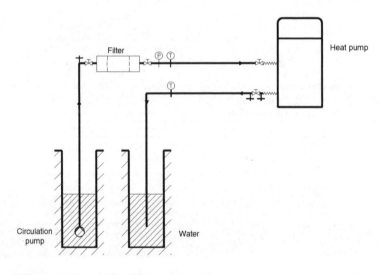

Figure 14.8 Direct use of rejected heat/heat source or sink
Source: Ochsner

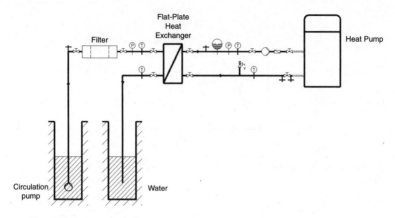

Figure 14.9 Indirect use of rejected heat/heat source or sink
Source: Ochsner

14.8 CO_2-heat pipes

A rare type of ground collector is the CO_2 heat pipe, which consists of a stainless steel pipe filled with CO_2 as heat transfer fluid. Other piping materials are also used.

Using the 'heat-pipe' concept, the fluid CO_2 runs down the inner side of the pipe, where it absorbs energy from the surroundings and evaporates and returns to the top of the pipe. This process occurs without the use of auxiliary energy or a pump. (Patent Prof. H. Kruse, FKW Hannover.)

At the top of the pipe, the CO_2 condenses, giving off the energy to the heat pump refrigerant.

The maximum depth of this patented method is approximately 100 metres. The heat collector capacity is comparable with that of a conventional brine vertical loop collector. Possible groundwater impacts are negligible.

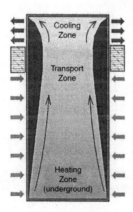

Figure 14.10 Heat pipe principle
Source: FKW

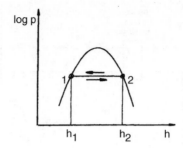

Figure 14.11 Log p - diagram CO$_2$ heat pipe
Source: Ochsner

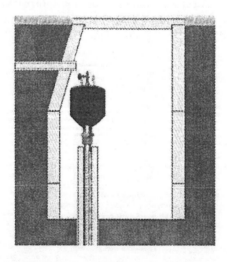

Figure 14.12 Construction of CO$_2$ heat pipe in a standard vault
Source: FKW

Figure 14.13 Construction of a borehole for the insertion of a CO_2 corrugated pipe
Source: Ochsner

15

Appendices

Appendix A Hydraulic schematics

Golf compact heating system
System variation overview

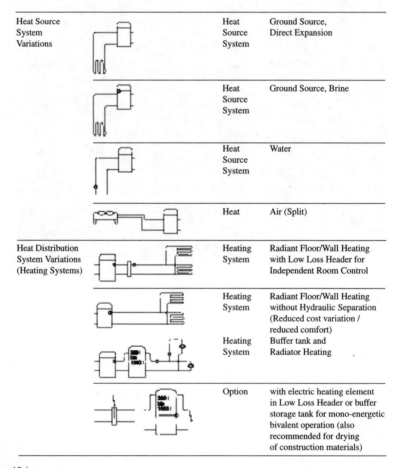

Heat Source System Variations		Heat Source System	Ground Source, Direct Expansion
		Heat Source System	Ground Source, Brine
		Heat Source System	Water
		Heat	Air (Split)
Heat Distribution System Variations (Heating Systems)		Heating System	Radiant Floor/Wall Heating with Low Loss Header for Independent Room Control
		Heating System	Radiant Floor/Wall Heating without Hydraulic Separation (Reduced cost variation / reduced comfort)
		Heating System	Buffer tank and Radiator Heating
		Option	with electric heating element in Low Loss Header or buffer storage tank for mono-energetic bivalent operation (also recommended for drying of construction materials)

Figure 15.1
Source: Ochsner

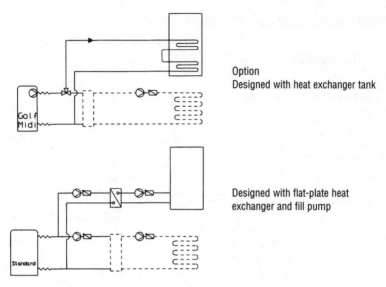

Option
Designed with heat exchanger tank

Designed with flat-plate heat exchanger and fill pump

Figure 15.2 Basic schematics for domestic hot water heating with heating heat pump and hot water tank
Source: Ochsner

HYDRAULIC SCHEMATICS – EXAMPLE

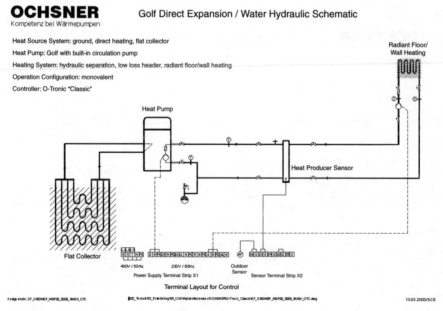

OCHSNER
Kompetenz bei Wärmepumpen

Golf Direct Expansion / Water Hydraulic Schematic

Heat Source System: ground, direct heating, flat collector
Heat Pump: Golf with built-in circulation pump
Heating System: hydraulic separation, low loss header, radiant floor/wall heating
Operation Configuration: monovalent
Controller: O-Tronic "Classic"

Radiant Floor/
Wall Heating

Heat Pump

Heat Producer Sensor

Flat Collector

400V / 50Hz   230V / 50Hz   Outdoor Sensor
Power Supply Terminal Strip X1   Sensor Terminal Strip X2

Terminal Layout for Control

13.03.2003/SCE

Figure 15.3
Source: Ochsner

HYDRAULIC SCHEMATICS – EXAMPLES

OCHSNER
Kompetenz bei Wärmepumpen

Ground / Water Hydraulic Schematic

Heat Source System: ground, direct expansion
Heat Pump: Golf with built-in circulation pump
Heating System: low loss header and electric element
Uses: heating/cooling
Domestic Hot Water: coil heat exchanger
Operation Configuration: monovalent
Controller: O-Tronic "Classic"

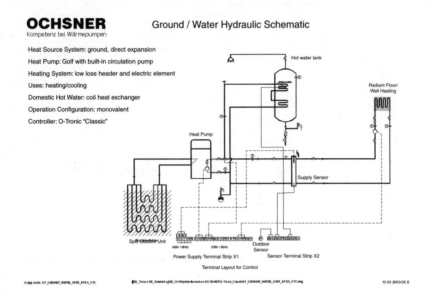

Figure 15.4
Source: Ochsner

OCHSNER
Kompetenz bei Wärmepumpen

Air / Water Hydraulic Schematic

Heat Source System: air, split outdoor unit
Heat Pump: Golf with built-in circulation pump
Heating System: separation/buffer tank, radiant floor/wall heating
Uses: heating/cooling
Domestic Hot Water: independent
Operation Configuration: bivalent, with electric heating element
Controller: O-Tronic "Classic"

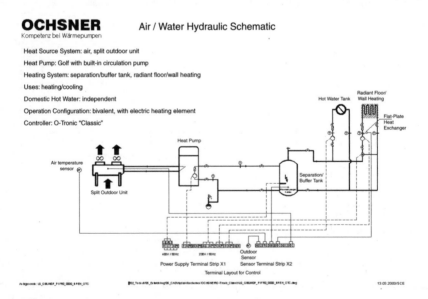

Figure 15.5
Source: Ochsner

HYDRAULIC SCHEMATICS – EXAMPLES

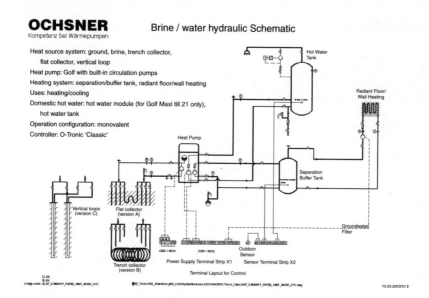

OCHSNER
Kompetenz bei Wärmepumpen

Brine / water hydraulic Schematic

Heat source system: ground, brine, trench collector,
    flat collector, vertical loop
Heat pump: Golf with built-in circulation pumps
Heating system: separation/buffer tank, radiant floor/wall heating
Uses: heating/cooling
Domestic hot water: hot water module (for Golf Maxi till 21 only),
    hot water tank
Operation configuration: monovalent
Controller: O-Tronic 'Classic'

Figure 15.6
Source: Ochsner

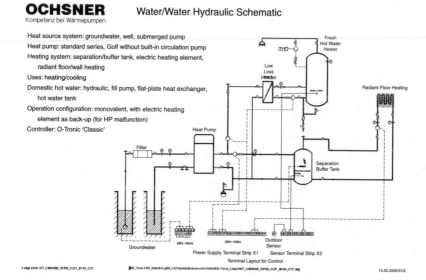

OCHSNER
Kompetenz bei Wärmepumpen

Water/Water Hydraulic Schematic

Heat source system: groundwater, well, submerged pump
Heat pump: standard series, Golf without built-in circulation pump
Heating system: separation/buffer tank, electric heating element,
    radiant floor/wall heating
Uses: heating/cooling
Domestic hot water: hydraulic, fill pump, flat-plate heat exchanger,
    hot water tank
Operation configuration: monovalent, with electric heating
    element as back-up (for HP malfunction)
Controller: O-Tronic 'Classic'

Figure 15.7
Source: Ochsner

HYDRAULIC SCHEMATICS – EXAMPLE

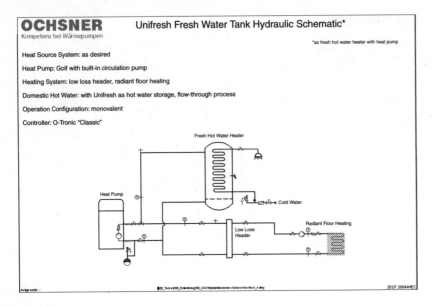

Figure 15.8
Source: Ochsner

A hydraulic connection of two heat pumps with buffer tank is shown in Figure 15.9.

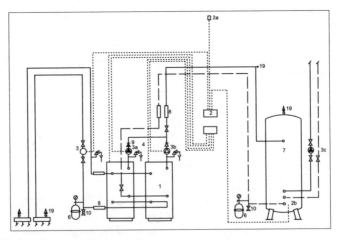

Figure 15.9 Hydraulic connection
Source: Stiebel Eltron

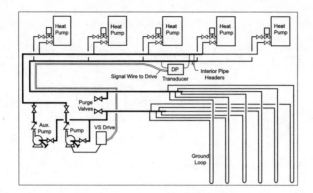

Figure 15.10 System with central heat source and decentralized heat pumps
Source: ASHRAE

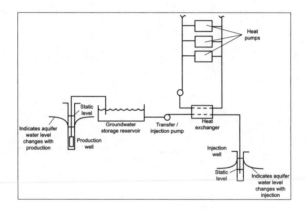

Figure 15.11 System with central heat source and decentralized heat pumps
Source: ASHRAE

Common Types of Heat Pumps

Heat Source and Sink	Distribution Fluid	Thermal Cycle	Diagram Heating Cooling Heating and Cooling
Air	Air	Refrigerant changeover	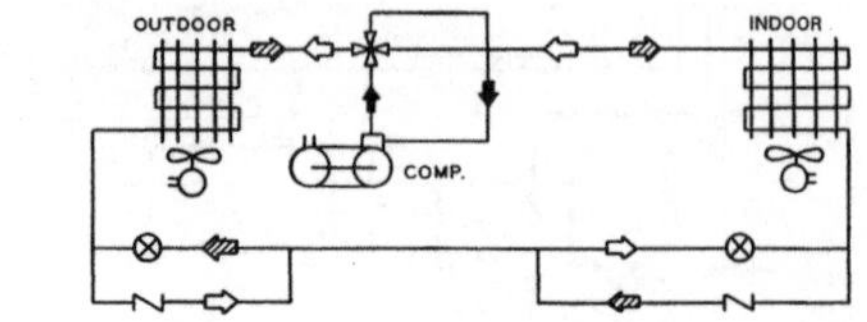
Water	Air	Refrigerant changeover	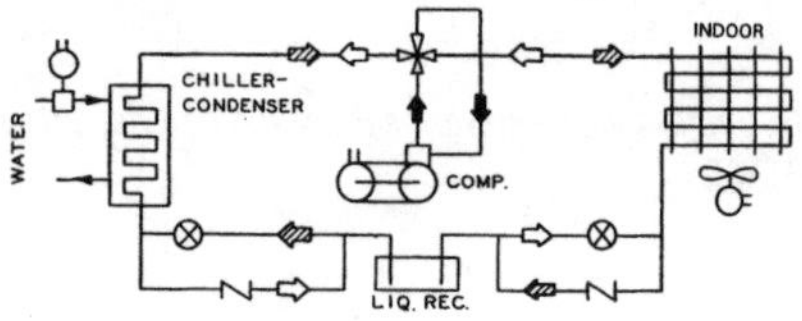
Water	Water	Water changeover	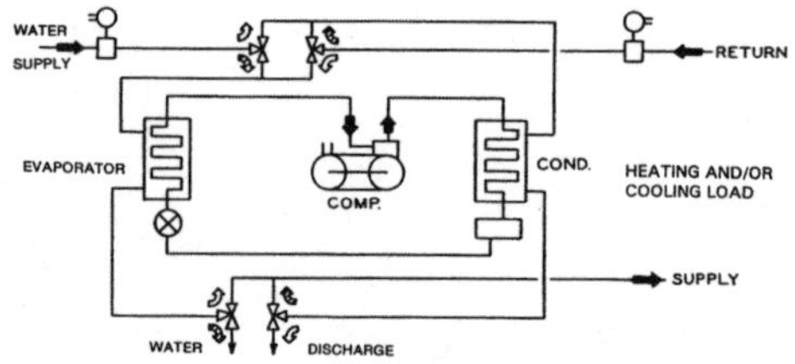
Ground-coupled (or Closed-loop ground-source)	Air	Refrigerant changeover	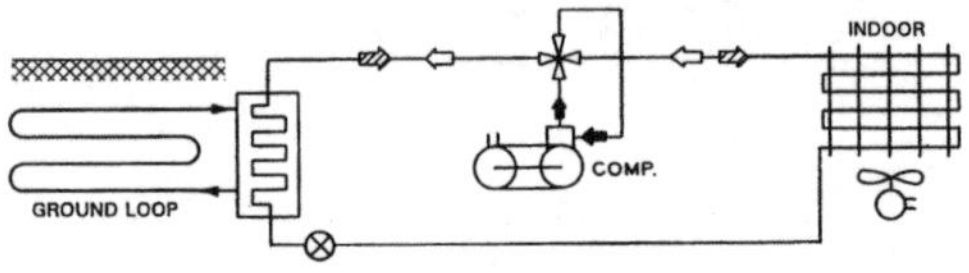
Ground-source, Direct-expansion	Air	Refrigerant changeover	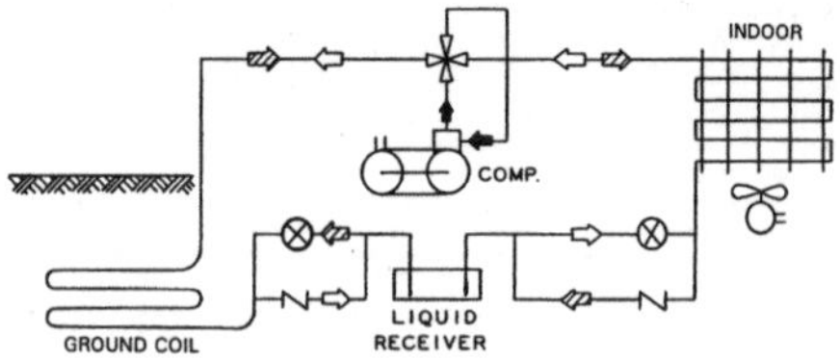

Figure 15.12 Heat Pump Systems
Source: ASHRAE, Handbook 2004

Appendix B Investment/operation costs (1/2007)

Investment Costs for New Building Heating with ca. 130 m² Living Area (D)
(Each comparable in comfort and quality!)

	Natural Gas Heating	Propane Heating	Oil Heating	Heat Pump
Cast boiler 9 kW, incl. weather driven digital controller	1,790 €	1,790 €		
Cast boiler with air injection oil burner 15 kW, incl. weather driven digital controller			3,300 €	
Ground source heat pump (GMDW) 9 kW incl. weather driven digital controller				5,900 €
Gas connection (complete), incl. tank for propane	1,800 €	1,900 €		
3000 liter oil tank (incl. accessories)			1,790 €	
Heat collector system (Ground Collector)				2,600 €
Mixer	350 €	350 €	350 €	
Mounting kits, safety devices, fixtures, accessories	850 €	850 €	850 €	450 €
Low-temperature radiant floor heating (insulation not incl.)	3,300 €	3,300 €	3,300 €	3,300 €
Heating loop distributor, including casing and room thermostat	700 €	700 €	700 €	700 €
Installation	1,400 €	1,400 €	1,400 €	1,400 €
Commissioning	210 €	210 €	160 €	650 €
Electrical connections	300 €	400 €	300 €	500 €
Cost of Heating System	*10,700 €*	*10,900 €*	*12,150 €*	*15,500 €*
Cost of Heating System with 20% Tax	*12,840 €*	*13,080 €*	*14,580 €*	*18,600 €*
Chimney	1,500 €	1,500 €	1,500 €	
Fuel tank half filled			330 €	
Tank room (incl. fire protection door)			1,700 €	
Excavation and ground preparation for collector installation				1,300 €
Sum of additional costs	*1,500 €*	*1,500 €*	*3,530 €*	*1,300 €*
Total cost incl. tax	*14,340 €*	*14,580 €*	*18,110 €*	*19,900 €*
Less subsidies (Government, Utilities)				

Figure 15.13 Investment cost comparison
Source: Ochsner

Operation Costs for New Building Heating with ca. 130 m² Living Area (D)

(130 m² x 50 W/m² -> 6.5kW x 1700 Hours of Operation per Year -> 11050 kW/a

	Natural Gas Heating		Propane Heating		Oil Heating		Heat Pump	
Energy demand (kWh)	11050		11050		11050		11050	
Average annual efficiency (%)	90		90		85			
Average annual working coefficient							4.5	
Heating value (kW/m3) (kW/l)	10.5		6.3		10			
Annual fuel consumption(m3) (l)	1169	m³	1949	l	1300	l		
Energy consumption (kWh)	12278						2456	
Price per unit quantity (€/m3) (€/l)			0.4		0.592			
Price per kWh (€/kWh)	0.051						0.11	
Energy cost per year		625 €		780 €		770 €		270 €
(=fuel consumption x fuel cost)								
Chimney sweep	45 €		45 €		45 €			
Interest ½ tank filling			27 €		27 €			
Maintenance costs		160 €		160 €		180 €		
Base price, taxes, metering cost	15 €	184 €						57 €
Sum of additional costs		389 €		232 €		252 €		57 €
Annual Operation Costs		1,014 €		1,012 €		1,022 €		327 €
Tax	20%	203 €	20%	202 €	20%	204 €	20%	65 €
Annual Operation Cost incl. Tax		1,217 €		1,214 €		1,226 €		392 €

Figure 15.14 Operation cost comparison
Source: Ochsner

ECONOMIC AND EMISSION COMPARISON
HEAT PUMP - COOLING MODE

Example: Single family house 220m², Heating demand 11 kW

1. Heat Pump GMSW 15 H/K (Brine/Water), Heating Capacity 11 kW

Cooling capacity, Climate Control heat pump (in cooling operation)		17.7 kW
Hours of cooling operation	60 d x 10 h =	600 h
Electricity demand for cooling operation (for EER = 5.4*)	600 h x 3.3 kW =	1980 kWh
Operation cost per summer (HP-rate)	1980 kWh x € 0.09 =	**178,2** €

Additional purchase cost for reversable mode of pump
(without the additional fan-coils required)   1,445.0 €

2. Air Conditioning conventional (air/air)

Electricity demand for cooling operation (for EER 2.6)		4,038 kWh
Operation cost per summer	4,038 kWh x € 0.15 =	**606.0** €

Cost for air conditioning (higher noise emission)
3 Window air conditioning units at 5 kW cooling capacity   3,600.0 €
Construction changes   500.0 €
Installation   300.0 €
                                                               4,400.0 €

Alternative
Single duct air conditioning unit 15 kW cooling capacity
(low noise emission)   6,400.0 €

3. Direct Comparison (Annual operation costs / Annual CO2 Emissions)

Annual Hours of Operation	Operation Costs			CO2 Emissions					
	Ochsner GMSW 15 H/K	Air Conditioning Air/Air	Savings CO2	Ochsner GMSW 15 H/K		Air Conditioning Air/Air		Savings € CO2	
				A	D	A	D	A	D
60 d	178.- €	606.- €	428.- €	546.3	1,169.1	1,198.6	2,565.0	652.3	1.395.9 kg
120 d	356.- €	1,212.- €	856.- €	1,092.4	2,338.2	2,397.1	5,130.0	1.304.7	2.791.8 kg
180 d	534.- €	1,818.- €	1,284.- €	1.638,7	3,507.3	3,595.7	7,695.0	1,957.0	4,187.7 kg

Additionally, it should be noted that during summer heat waves the local grid is often overloaded, which is some locations leads to power outages.

*including circulating pumps

Figure 15.15 Economic and emission comparison heat pump cooling mode

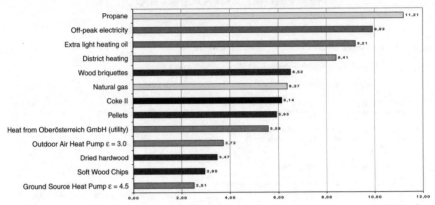

1 kWh costs ..... Cents

Propane	11,21
Off-peak electricity	9,92
Extra light heating oil	9,21
District heating	8,41
Wood briquettes	6,52
Natural gas	6,37
Coke II	6,14
Pellets	5,93
Heat from Oberösterreich GmbH (utility)	5,58
Outdoor Air Heat Pump ε = 3.0	3,72
Dried hardwood	3,47
Soft Wood Chips	2,95
Ground Source Heat Pump ε = 4.5	2,51

Price of 1 kWh of useful heating energy = gross fuel price WITHOUT related costs such as electric demand, service, fuel deposit, etc. ...

Actual Prices from 1 September 2006
Values in Cents

Figure 15.16
Source: Energie AG

SYSTEM DATA SHEET **OCHSNER**
Heat Pump Heating System

The system data sheet enables the pre-initialization of your heat pumps control as well as the optimization according to the instal-lation conditions. For this reason, your heat pump is completed after the system data sheet is returned. Furthermore, it is a requi-rement for the performance guarantee.

Date:	Your Order:			Our Work Order			
	Installer / System Partner			System Operator / End Customer			
Issued by:							
Heat Pump Model:			Operation:		Heating		Cooling

HEATING SYSTEM

Operation Configuration		Heat Distibution			Control		
	Monovalent		Radiators			No Controller	
	Bivalent Parallel		Radiant Floor/Wall			O-Tronic "Classic"	
	Bivalent Alternate		°C	Auxiliary Set Point		O-Tronic "Classic" Plus	
	Electrical Element as Back-Up				→	Auxiliary Heater	
Power Supply					→	Legionella Function	
	Voltage		1/3 Phase		50/60 Hz	→	Error Control for Master Building Controller Gebäudeleitsystem
Total Heating Capacity		kW	Hydraulic Separation			Other	
at Standard Temperature		°C	Low-Flow Header			Remote Control	
Planned Supply Temperature		°C	Buffer Tank				Remote Control with Land-Line Telephone
Supply/Return Piping Diameter			No Separation				Remote Control with GSM -Cellular Telephone
Mixing Loop			Only without zone or mix control and without thermostat valve possible!				Communication Pack (analog modem)
No mixer							Communication Pack (direct PC-connection)
Number of mixers			max. 7			Room Remote Control	
Domestic Hot Water Heating with Heat Pump							No Remote Control
yes	with hot water module in fill tank principle		with Unifresth				RC with Sliding Switch
	with circulator pump		with insterted heat				RC with LCD- Display
	with 3 - way valve		with coil in tank (for Golf Midi up to 6 kW)				
	electrical element for domestic hot water		with flat- plate heat exchanger				
no	with Europa Model						

HEATCOLLECTOR SYSTEM

Brine							
	Flat Collector		Number of Loops		m²	Installation Area	
	Trench Collector		Number of Loops		m	Total Trench Length	
	Earth Tap		Number of Taps		m	Drilling Depth	
Collection Trench	yes	no			m	Connection Length	
Direct Heating			Number of Loops		m²	Installation Area	
Collection Trench	yes	no			m	Connection Length	
External Expansion Valve	yes	no					
Air					m	Connection Length	
Water							
Water Filter Type			Water Analyses Prepared	yes		no	
Min. Cource Temp.	°C		Water Analyses Attached	yes		no	
Continuous Pump Trial	m³/h		Gorund Water Depth		m		
Date/Signature Installer/System Partner		Accepted by			Date Inspected / Signature Sales Service		

Permission may be required for all ground or groundwater source system. All heating system hydraulic and electrical connection must be completed before comissioning. The heat pump must be installed according Ochsner guidelines to ensure the performance.

Figure 15.17 Sample of a system data sheet for heat pump heating system

C448.2-02   © *Canadian Standards Association*

Annex A (Informative)
Installation Checklist for Open- and Closed-Loop Earth Energy Heat Pump Systems

Note: *This Annex is not a mandatory part of this Standard.*

(Two Copies Are to Be Provided to the Owner)

Owner's Name _____ Date _____
Address _____
Province _____ Postal Code _____ Phone _____
Contractor's Name _____ Date _____
Address _____
Province _____ Postal Code _____ Phone _____
System Type: Open-Loop ☐   Closed-Loop ☐   House Type _____
Design Heat Load (Building) _____ Design Method _____
Design Cooling Load _____ Method _____
Domestic Hot Water Load (Met By System) _____
Total Heating Load _____
Type Of Distribution System: Forced-Air ☐   Hydronic ☐
Heat Pump Make _____ Model/Serial No. _____
Heating Capacity _____ Cooling Capacity _____
Check off appropriate entering water temperatures   Heating EWT: 0°C (32°F) ☐   10°C (50°F) ☐
(EWT). (Refer to CSA Standard CAN/CSA-C13256-1)   Cooling EWT: 25°C (77°F) ☐   10°C (50°F) ☐

If A Closed-Loop System:

Heat Exchanger Length, if Horizontal _____
Heat Exchanger Type, if Horizontal   Single-Pipe ☐   Two-Pipe ☐
                                      Four-Pipe ☐   Other ☐
Borehole Depth and Number, if Vertical _____
Heat Exchanger Sized According to: Manufacturer ☐

If Software, Program Used: _____
Backfill Materials, Horizontal Trenches _____
Borehole Fill Material, if Vertical _____
Type Of Antifreeze/Inhibitors _____ Quantity _____
Antifreeze Protection Level _____ Loop Test Pressure _____
System Static Pressure _____

If An Open-Loop System:
Attach copy of water well record or well pump test and include the number and specifications of wells, intake, and pumps.

Marking/Instructions Checklist

If A Closed-Loop System:

Supply and Return Valves Marked Accordingly	☐
Submerged Heat Exchanger Position Marked at Shoreline	☐
Label at Loop Charging Valve Showing Antifreeze Type, Concentration, Contractor Information	☐
Owner Given Manufacturer Documentation and Warranty on System	☐
Owner Given Site Survey Worksheet of Installed System (Including Dimensions/Locations of all Piping, Diameter, Depths and Lengths of Loops, Septic Systems, Water Inlet Lines, Lot Lines, etc.)	☐

If An Open-Loop System:

Supply and Return Lines to be Identified by Marker at Point of Entry to Water Wells	☐
Inform Owner of Possible Effects on Supply Water Well of Open-Loop System — Water Quality, Quantity, etc.	☐
Ensure Water Supply Well is Sealed in Accordance with Approved Well Construction Practices	☐
Ensure Water Well Yields Water to Supply Both Domestic and Heat Pump Requirements at Time of Installation	☐

This installation was done in accordance with CSA Standard C448.2, *Design and Installation of Earth Energy Systems for Residential and Other Small Buildings,* and currently applicable regulations.

Name: (Please Print or Type) _____ Signature _____
Date _____

February 2002

Figure 15.18 Sample of installation checklist for open- and closed-loop earth energy heat pump systems

Test conditions for the determination of heating capacity

Water-loop heat pumps	Ground-water	Ground-loop	
heat pumps			
heat pumps			
Liquid entering indoor side	40°C	40°C	40°C
Air surrounding unit, dry bulb	15°C to 30°C	15°C to 30°C	15°C to 30°C
Standard rating test	20°C	10°C	0°C
Liquid entering outdoor-side heat exchanger			
Part-load rating test	20°C	10°C	5°C
Liquid entering outdoor-side heat exchanger			
Frequency[a]	Rated	Rated	Rated
Voltage[b]	Rated	Rated	Rated

[a]Equipment with dual-rated frequencies shall be tested at each frequency.
[b]Equipment with dual-rated voltages shall be performed at both voltages or at the lower of the two voltages if only a single rating is published.

Figure 15.19 Test conditions ANSI Standard
Source: ANSI/ARI/ASHRAE ISO Standard 13256-2:1998

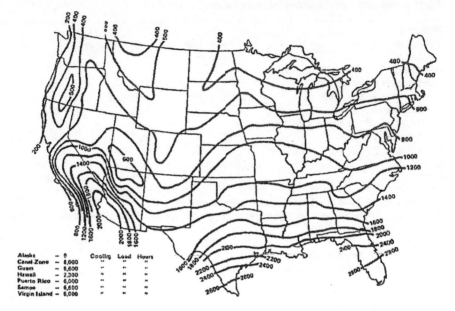

Distribution of actual cooling load hours (CLH$_a$) throughout the United States.

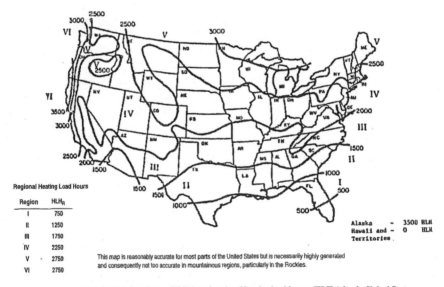

Region	HLH$_R$
I	750
II	1250
III	1750
IV	2250
V	2750
VI	2750

This map is reasonably accurate for most parts of the United States but is necessarily highly generated and consequently not too accurate in mountainous regions, particularly in the Rockies.

Actual heating load hours (HLH$_a$) and regional heating load hours (HLH$_R$) for the United States.

Figure 15.20 US climate zones
Source: ANSI/ASHRAE Standard 116-1995/RA 2005

Appendix C Units, symbols and conversion

Quantity	Symbol	Designation	Unit	Additional units (definition)
Weight	m	Kilogram	kg	
Density	ρ	Kilogram per cubic meter	kg/m³	
Time	T	Seconds	s	1 h = 3600 s
		Hours	h	
Flow rate	$\dot{V}$	Cubic metre per second	m³/s	
Mass flow rate	$\dot{m}$	Kilogram per second	kg/s	
Force	F	Newton	N	1 N = 1 kg m/s²
Pressure	P	Pascal	N/m², Pa	1 Pa = 1 N/m²
				1 bar = 105 Pa
Energy, Work, Heat (quantity)	E, Q	Joule	J	1 J = 1 Nm = 1 Ws = 1 kg m²/s²
		Kilowatt-hour	kWh	1 kWh = 3600 kJ = 3.6 MJ
Enthalpy	H	Joule	J	
Heat Transfer Rate	P, Q	Watt	W	1 W = 1 J/s = 1 Nm/s
		Kilowatt	kW	(work per time, i.e. heating demand)
Temperature	t, T	Kelvin	K	Absolute, temperature difference
		Celsius	°C	Temperature in Celsius
Entropy	S	Joule per Kelvin	J/K	DIN 1345 (1.72)
Thermal conductivity	λ	Watt per Kelvin-Meter	W/(Km)	DIN 1341 (11.71)
Heat transfer coefficient	κ	Watt per Kelvin-m²	W/(Km²)	DIN 1341 (11.71)
Sound level Sound pressure	L	Decibel	dBA	Sound power level
Efficiency	η			Efficiency (output/input)
Coefficient of performance	ε (COP)			Performance figure (Heating capacity/input power)
Average efficiency	β			Annual working coefficient
Specific heat capacity	c	Joule per Kelvin-kg	J/(K kg)	
	ODP	Ozone depletion potential	CO_2	Ozone layer destruction potential of a gas
	TEWI	Total equivalent Warming impact	CO^2	Total greenhouse potential

FUELS
Heating values, CO_2 emission

Fuel	Lower Heating Value* (H_u)		Higher Heating Value** (H_o)		Max CO_2 Emission (kg/kWh) based upon:	
					LHV – H_u	HHV – H_o
Hard Coal	8.14	kWh/kg	8.41	kWh/kg	0.350	0.339
Heating Oil EL	10.08	kWh/l	10.57	kWh/l	0.312	0.298
Heating Oil H	10.61	kWh/l	11.27	kWh/l	0.290	0.273
Natural Gas L	8.87	kWh/m$_n^3$	9.76	kWh/m$_n^3$	0.200	0.182
Natural Gas H	10.42	kWh/m$_n^3$	11.42	kWh/m$_n^3$	0.200	0.182
Propane	12.90	kWh/kg	14.00	kWh/kg	0.240	0.220

* The Lower Heating Value, H_u, is the quantity of heat released during the full combustion of one cubic metre gas at standard conditions (0°C, 1013.25 bar) if the energy content of the water vapour goes unused, as is the case in most all residential, agricultural, commercial and industrial heating systems.
** The Higher Heating Value, H_o, is the quantity of heat released during the full combustion of one cubic metre gas at standard conditions (0°C, 1013.25 bar) if the heat of vaporization is recovered through condensation of the water vapour. The HHV H_o is used in few specialized gas applications. For consumption calculations with gas utilities, the LHV H_u is generally used.

US and British units & conversion factors

Flow rates

1 m³/h	= 4.403 gal/min (US)	1 gal/min (US)	= 0.227 m³/h
(fluids)	= 3.666 gal/min (UK)	1 gal/min (UK)	= 0.273 m³/h
1 m³/h (gases)	= 0.5883 ft³/min	1 ft³/min	= 28.317 l/min

Pressure

1 bar	= 14.504 psi (lb/in²)	1 psi (lb/in²)	= 68.95 mbar

Energy

1 kWh	= 3414.5 Btu	1 Btu	= 1.055 kJ

Power

1 kW	= 3412 Btu/h	1 Btu/h	= 0.2931 W

Refrigeration Capacity

1 kW	= 0.2843 tons	1 ton	= 3.517 kW

Temperature

$\dfrac{°F - 32}{1.8}$	= °C	°F = (°C x 1.8) + 32	
0°C	= 32°F	50°C = 122°F	

Source: Ochsner

CENTRAL HEATING PIPE SIZING
in Function of Flow Rate

Flow Resistance Diagram for Steel Pipes

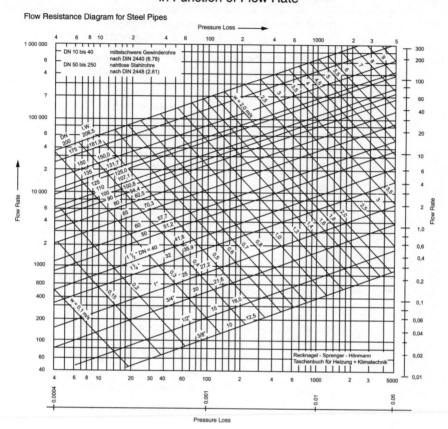

Figure 15.21 Friction Loss Diagram for Steel Pipes
Source: Recknagel

Index of Figures and Sources

Figure	Source	Address
1.1	Prof. D. Schönwiese	Institute for Geophysics, University Frankfurt/M
1.2	Energy Report for the Austrian National Government	Institute for Thermal Technic TU Graz. Prof. H. Halozan A 8010 Graz
1.4, 8.11	AWP-Zürich	Arbeitsgemeinschaft Wärmepumpen CH 8023 Zürich
1.5, 13.1, 15.16	OKA, Energie AG	Energie AG OÖ A 4021 Linz
1.6	BGR	Bundesanstalt für Geowissenschaften u. Rohstoffe D 30655 Hannover
1.9	UN-Population Report 97	UN-Population Report 97
1.3, 1.10, 1.11	VDEW	Verein deutscher Elektrizitätswerke 60329 Frankfurt
2.1, 9.10	Building Advice Guide 1998	RIP Media Periodicals and Publishing GmbH: 1998, Vienna.
3.15	Thermo Energie GmbH	Thermo-Energie Systemtechnik GmbH A 3350 Stadt Haag
3.18, 13.2	Helios	Helios Ventilatoren GmbH & Co D 78056 Villingen-Schwenningen
4.1, 4.2, 9.11	DIN	Deutsche Institute für Normen e.V.D. 50672 Köln
2.2, 2.5, 4.3, 4.7	BWP	Bundesverband Wärmepumpe D 80796 München
5.4	Prof. B. Sanner	Justus-Liebig University Gießen Institute für Angew. Geowissenschaft D 35390 Gießen
1.8	WTRG Economics	
2.7	Copeland	Copeland GmbH D 13509 Berlin
3.3, 3.8, 3.12, 3.13	AWP	AWP Wärmepumpen GmbH D 99310 Arnstadt
9.3, 15.21	Recknagel	Recknagel, Sprenger, Schamek: Taschenbuch für Heizung, Klima, Technik, 1995, München.

Index of Figures and Sources

Index of Tables and Sources

Table	Source	Address
5.1, 5.3, 6.1, 13.1	VDI 4640	Verein Deutscher Ingenieure D 40002 Düsseldorf
12.1, 12.2, 12.3	VDEW	Verein deutscher Elektrizitätswerke 60329 Frankfurt
5.2, 7.1, 14.1	Ochsner	Ochsner Wärmepumpen GmbH...

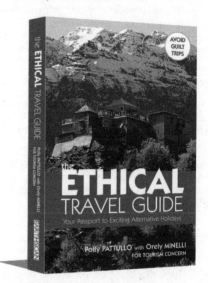

Join our
online community
and help us save paper and postage!

www.earthscan.co.uk

By joining the Earthscan website, our readers can benefit from a range of exciting new services and exclusive offers. You can also receive e-alerts and e-newsletters packed with information about our new books, forthcoming events, special offers, invitations to book launches, discussion forums and membership news. Help us to reduce our environmental impact by joining the Earthscan online community!

How? – Become a member in seconds!

>> Simply visit **www.earthscan.co.uk** and add your name and email address to the sign-up box in the top left of the screen – You're now a member!

>> With your new member's page, you can subscribe to our monthly **e-newsletter** and/or choose **e-alerts** in your chosen subjects of interest – you control the amount of mail you receive and can unsubscribe yourself

Why? – Membership benefits

✔ Membership is free!

✔ 10% discount on all books online

✔ Receive invitations to high-profile book launch events at the BT Tower, London Review of Books Bookshop, the Africa Centre and other exciting venues

✔ Receive e-newsletters and e-alerts delivered directly to your inbox, keeping you informed but not costing the Earth – you can also forward to friends and colleagues

✔ Create your own discussion topics and get engaged in online debates taking place in our new online Forum

✔ Receive special offers on our books as well as on products and services from our partners such as _The Ecologist, The Civic Trust_ and more

✔ Academics – request inspection copies

✔ Journalists – subscribe to advance information e-alerts on upcoming titles and reply to receive a press copy upon publication – write to info@earthscan.co.uk for more information about this service

✔ Authors – keep up to date with the latest publications in your field

✔ NGOs – open an NGO Account with us and qualify for special discounts

Join now?
Join Earthscan now!
name
surname
email address

Earthscan Member
[Your name]

Click to Change

My profile
My forum
My bookmarks
All my pages

www.earthscan.co.uk

2647